This book was made possible by funding from the Open Society Foundation for South Africa.

The views expressed in this work are those of the author and do not necessarily reflect those of the Open Society Foundation for South Africa.

The Unwritten Rules of Policing South Africa

JONNY STEINBERG

JONATHAN BALL PUBLISHERS
Johannesburg & Cape Town

with
OPEN SOCIETY FOUNDATION FOR SOUTH AFRICA

Published in 2008 by
JONATHAN BALL PUBLISHERS (PTY) LTD
P O Box 33977
Jeppestown 2043

ISBN 978 1 86842 303 3

The author has asserted his moral rights.

Cover design by Lucky fish, Durban
Design and typesetting by Etienne van Duyker, Cape Town
Printed and bound by CTP Book Printers, Parow, Cape

Acknowledgments

My primary debt is to the Open Society Foundation of Southern Africa (OSFSA), which funded this project, and, in particular, to the former director of its Criminal Justice Initiative, Sean Tate, and its executive director, Zohra Dawood. Sean approached me in early 2007 with an idea for a short book on policing; he and Zohra were happy to humour my vaguely formulated ideas on what the book should look like, and gave me the space to develop them. Many thanks for that. My thanks, also, to Louise Ehlers, Sean's successor at OSFSA.

I wrote this book while a visiting fellow at the African Studies Centre at the University of Oxford. I am grateful to the Ernest Oppenheimer Fund Committee, which administers the fellowship. My warm thanks too to David Anderson, Nic Cheeseman, and William Beinart, who made me feel at home at the African Studies Centre and went out of their way to give me opportunities to present my work to as wide and varied an audience as possible.

I am grateful to Jean-Paul Brodeur at the University of Montreal, who showed me draft chapters of a work in progress, provisionally titled *A Treatise on Policing*, to be published by University of Toronto Press. Reading his work helped me a great deal to interpret what I had seen of South African policing from the back of police vans.

Antony Altbeker and Alex Dodd both read drafts of the manuscript and provided invaluable commentary. Jane Rogers at Jonathan Ball Publishers did a fine job editing it. And thank you, once again, to Jonathan Ball, Jeremy Boraine, Francine Blum and Tanya White, who, over the years, have taken such fine care in the production and marketing of my books.

Jonny Steinberg
May 2008

Contents

Two incidents, juxtaposed

Consider these two incidents. They took place on the same day, in the same town, and involved the same two police officers.

The town was the old mining settlement of Randfontein, about 40 kilometres west of downtown Johannesburg. The policing sector was Toekomsrus, Randfontein's coloured township. And the shift was Saturday night – 6 pm to 6 am – one of the busiest and most difficult times for township cops, for about a third of their weekly crime load is packed into these 12 hours.

Toekomsrus is home to some 30 000 people. It is one of those inward-turned spaces the apartheid government built for black and coloured people in the early 1970s; concentric layers of crescent-shaped streets, the outermost layer, Diamant Street, forming a closed circle around the periphery of the township.

Deep within Toekomsrus's inner layers are two shebeens. By nine or ten o'clock on a Saturday night, each has attracted a clientele of a good 500 people, few of them older than 30. They spill out onto the street in dense clusters, blocking passing traffic, making the public space around the shebeen their own.

Soon after the 6 pm shift began, the officers with whom I was riding along, Constables K and N, attended to a minor traffic accident on a regional artery that nudges the southwestern corner of their jurisdiction. There was some friction between the two parties involved, but nothing that required a police officer's intervention. Constables K and N's role was mechanical and bureaucratic; they did the necessary paperwork, issued advice on claiming for insurance, waited until both parties had left the scene, and then got into their van and drove off. The radio was silent for the better part of the next hour. We circled the township over and again, tracing a wide arc around it, seldom venturing much deeper into Toekomsrus than Diamant Street.

Constables K and N had a new device in their car. When they found themselves within five kilometres of a vehicle whose satellite tracker had been activated, signalling that it had just been stolen, the vehicle's number plate flashed on a small monitor on their dashboard. A bar on the side of the screen told them how far they were from the vehicle.

At about 7.30 pm, the constables' monitor flashed a number plate at them. Both immediately snapped out of the lethargy of their early Saturday evening routine. Constable N sat up ramrod straight in the passenger seat. Constable K began fiddling with the display on the vehicle tracer, as if it were a poorly tuned radio. Ostensibly, the constables were minutes away from pursuing a car full of armed men.

They followed the signal. Doing so was a process of trial and error: it required a lot of reversing and turning around; you drive, you watch the bar on the side of the monitor, you see whether it is getting longer or shorter.

Within about 15 minutes, the vehicle tracer had put

us on a road on the outskirts of Randfontein, the patrol van's nose pointing into an open field. On the other side of the field, perhaps 250 metres away, was the Soweto car pound, temporary home to several hundred scrapped motor vehicles. The bar on the side of the screen told us that the vehicle in question was 250 metres away. It could not possibly have been to the left or the right of us, nor behind, since we had explored in all of those directions already. The car we were looking for was without question directly in front of us, in the pound. The tracking service had surely forgotten to turn the signal off. It was a false alarm.

Constables K and N were undeterred. In their two-wheel-drive sedan, they mounted the pavement and drove into the darkness of the field. As we blundered along, our heads hitting the ceiling of the patrol van, its headlights pointing now at the treetops, now at the ground, what the constables were doing became abundantly clear: they were avoiding the inner rings of Toekomsrus.

Just over halfway through the shift, I discovered why.

Having been dispatched to three domestic violence complaints during the course of the night, a call of a different order came over the radio at about 12.30 am. A man had phoned the station to report that his son had taken his bakkie without his permission, that the vehicle had been spotted outside one of the township's two shebeens, and that he wanted it back.

If Constables K and N felt fear or discomfort at this news, they did not show it. They drove to the shebeen, nudged the drinkers in the street out of their way with their front bumper, spotted the bakkie that was the subject of the complaint, and parked directly behind it. They then walked into the shebeen, and returned a few minutes later with the young man in question.

The crowds of shebeen patrons minded their business all the while. The cops seemed invisible to them. They were scattered across the street and on the pavements and in the shebeen yard in dense clumps, like a massive, rowdy meeting that had broken into small groups to discuss a matter at hand.

The constables got back into their van and waited for the youngster to get into his father's bakkie. He didn't. He stood in a huddle with some other young men; their conference lasted quite a while. Constable K hit the patrol van's hooter in irritation, and the huddle immediately dispersed; half a dozen youngsters climbed into the back of the bakkie, the last one into the driver's seat. They drove off, and we followed.

A few hundred metres from the shebeen, the bakkie turned into a vacant field and stopped. The constables followed, then flashed their lights, then hooted. The bakkie remained stationary, and the young men remained on the back, absorbed in a loud and voluble conversation, as if the constables were not there at all.

K and N got out of their van and walked towards the bakkie. The youngsters kept ignoring them. And then the constables crossed some invisible line between their own vehicle and the bakkie, and the young men sprang to life, nimble and fast, like a six-bodied machine awoken by a switch, and within seconds they had left the bakkie and arranged themselves into a tight, menacing circle around the cops.

One of them pointed at Constable K and began to

scream. He was no more than 5ft 6in and quite slight, but he stood there with feet wide apart, his mouth just a few inches from Constable K's, and his confidence gave him all the stature he needed. His nostrils flared, his eyes bulged, and his pelvis thrust itself intermittently at Constable K, as if his body willed violence, only his self-discipline holding him in check. He spoke very fast in a colloquial Afrikaans to which I was not accustomed, and it took some time for it to dawn on me that he was threatening to kill Constable K's children. He named them both, named their primary school, shouted out that he knew school ended at twenty past one every week day except Wednesdays, when it ended at quarter to one.

I lost my bearings: the shock of this sudden assault disorientated me, and for some time I was groundless, unable to read the meaning of the situation, or to predict the course it might take. And then it came to me slowly that the encounter was not emitting enough energy, not as much as it should have were it truly unstable. In retrospect, I would understand that it was in fact tightly choreographed, that all of its participants knew precisely how it would proceed and how it would end. The central question was one of numbers, and an anticipation of how the balance of numbers was set to change in the minutes to follow.

Constables K and N were two against seven. If they tried to arrest anybody, they would be overpowered, their guns taken from them. Nor could they get into their vehicle and drive away: an attempt to retreat would surely enrage the youths, and incite them to violence. They were trapped into standing there and listening to the short one volley death threats at Constable K's children. The only way to get out of this trap was to increase their own number, to get more of themselves onto the scene. The constables proceeded to do precisely that. They backed away carefully towards their car. Constable N leaned in through the window and radioed for help. He then put the receiver back in the car, stood up straight, and stared at his screaming assailants.

The youths knew that they had three or four minutes left, and they made the most of their shrinking time, volleying gorier and more vivid death threats. Constable K had now retreated into the car and was sitting in the passenger seat, the window wide open. Making the most of his unaccustomed advantage in height, K's tormentor stood at the window jabbing his finger into K's chest, the only moment of physical contact throughout the encounter that I can recall.

At the sound of the approaching sirens, the youths tensed. Their shouts and insults grew more furious. They grew lighter on their feet. They waited, and waited, coiled and ready, until two blaring patrol vehicles mounted the sides of the field at opposite corners. And then they bolted, lean and fast, each in a different direction, into the night. The moment they turned to run, Constables K and N leapt from their vehicle and pursued them, but they were sluggish and potbellied, and from the back of the patrol car I watched them lumbering around like clowns.

*

Constables K and N spent the next three hours in search of the boy who had threatened to kill K's children. His name, I had learned by now, was Darren, and K had arrested him six months ago; he had spent three of those months in prison. What I had witnessed tonight was a gift for Darren: an unexpected opportunity for revenge.

At about 2 am we found Darren, alone, standing in our headlights about a hundred paces in front of us, his fingers in his pockets, his narrow body throwing a long shadow towards us. Constable N got out of the car and drew his gun. Darren stood still. Constable N stepped away from the car, into the darkness where Darren could not see where his gun was pointing, and fired two rounds into the air. The gunshots cracked the roof of the night sky and echoed back at us. My first thought was that they could be heard all over Toekomsrus; I wondered how many imaginations had in that instant conjured a different story to explain the gunshots; a record of all those stories, I found myself thinking, would probably document every fear this place has of itself and its young men.

Darren still stood there, motionless.

Defeated, Constable N leapt back into the car, and we sped towards Darren, but by the time we got to the place he'd been standing, he was long gone, and although the constables leapt out of their van and made a show of searching the surrounding gardens and yards, they knew they would not find him.

We drove off in silence, and the silence lasted a long time. It was hard not to admire Darren. His performance was no outburst of drunken exuberance. It was quite efficient, quite eloquent. He understood with practical agility something which, I would come to see in time, all young township men understand: that the question of space is irrelevant, the question of numbers all important. When young men accumulate in a large group, the ground on which they stand is theirs; when they disperse, the space belongs, once again, to the police.

Duty had brought Constables K and N into a space filled with young men, a space they had been avoiding all night. As quick as a flash, Darren understood that he had in his hands a tactical weapon, a means to convey to K that he would pay dearly for that arrest of six months ago, that to police the young men of Toekomsrus comes at a heavy cost.

*

An hour or so later, we came across a group of young men and women walking home from the shebeen. Constable K rolled down the window and told them to tell Darren that one day when he wasn't expecting it, K would knock down Darren's door in the early hours of the morning, ram the barrel of his gun into Darren's mouth, and blow his brains out.

'I am going to kill him,' Constable K said softly. 'The only language Toekomsrus understands is the language of fear.'

The kids kept their heads down and walked.

*

And now the second incident. It was close to dawn, about five hours after the routing the two constables had received in the field.

K and N had an end-of-shift ritual. Fifteen minutes before they were due to knock off, they drove into the parking lot of a 24-hour takeaway joint in Randfontein's central business district. They each bought a polystyrene cup of milky instant coffee, returned to their car, drank it, then drove back to the station and went home to sleep. It was their winding down time.

The constables were sitting in their car drinking their coffee when an elderly white man walked towards their window, stopped at a respectable distance, and nodded at them cautiously. They looked at him, and he began to speak uncertainly and with careful diffidence. Constables K and N, by the way, are black Africans, the only two I had seen the entire evening.

The white man told them that he was a landlord, that he owned a flat just around the corner, that he had a tenant, a young black woman called Gladys, and that Gladys had not paid him rent in three months. Would they speak to her? As he waited for them to reply, it was clear that he could not begin to anticipate how they might respond, that he was preparing himself for anything from a scolding to being entirely ignored.

The officers glanced at one another, and then Constable K nodded his assent, saying that they would first finish their coffee. The elderly gentleman's face flushed with welcome surprise. He went to his car, an ancient blue Datsun, and waited.

Minutes later, we followed him to a three-storey block just a few hundred paces away. He pointed us to a tiny, freestanding room on the side of the grounds.

Constable N pounded on the door with the back of his heavy torch, then kicked it for good measure. Constable K shouted: 'Gladys, it's the police! Open the door!' It was now just before six o'clock on a Sunday morning.

We heard keys rattling, locks unlocking and bolts unbolting. A sleepy-eyed, very frightened young woman in a skimpy nightdress opened the door. The constables pushed past her into the room. A young man was sitting on the bed stark naked, his body tensed in self-defence, his hands searching for linen with which to cover himself. The constables both stared at him long enough to convey cheap contempt and then paid him no further attention. They turned to Gladys and spoke to her in Tswana for some time. I do not understand much Tswana, but the tone of their voices was not polite: it was of a piece with the manner in which they had announced their presence.

I went outside and waited with the landlord. He held his body very uncomfortably, his arms folded tightly across his chest in the morning's brisk cold and, as he regarded me nervously, unsure of precisely who I was and what I was doing with the uniformed men shouting at his tenant, his eyes revealed something exquisitely complex.

He was, on the one hand, lapping up the scene quite greedily: these two hefty men with their uniforms and guns and handcuffs, expressing his anger on his behalf, giving him a potency which ten minutes earlier he could scarcely have imagined. And yet he was also eyeing the future, anticipating what this incident would come to mean once it was over and the police had left. Then it would just be him and Gladys and her boyfriend and the memory of this terrible invasion of their privacy that these agents of his anger had inflicted. 'I have overreached myself,' his eyes seemed to say. 'I have borrowed a force too blunt for me to handle.'

*

'That was not police business,' I said cautiously from my place in the backseat on the way back to the station.

'You are correct,' Constable K replied.

'But it would have become police business,' Constable N chipped in, completing his colleague's sentence.

'It starts like this,' Constable K continued. 'A dispute over rent. Next thing they are hitting each other over the head with thick pieces of metal, and then we have to come and stand with the body and wait for the mortuary van. It's better that we intervene now, even though we are tired and it means we will get to our beds half an hour late.'

It struck me, as I listened to K, that his tone was that of a primary school teacher: you can see it coming from a mile, what these kids are going to do to each other; it is for us world-weary adults to intervene.

*

That was in June 2004. It is now, as I write, September 2007. In the intervening three years, I have spent about 350 hours riding along in the patrol vans of the South African Police Service. I think of that night in 2004 often. I have come to see the juxtaposition of the two incidents that punctuated that shift, the routing of Constables K and N at midnight, and their routing of Gladys at daybreak, as a pretty good analogue of the trajectory policing has taken in post-apartheid South Africa.

The most important precondition for policing in a democratic society is the consent of the general population to be policed. A people that is policed is one that lives in a condition of civil peace. Collectively, it understands that disputes and conflicts do arise, but it regards these as temporary ruptures, not as threats to the underlying order. It accepts that there are men who beat their lovers and brawl in bars, tenants and landlords who come to blows, that some people try to earn a living holding up others at gunpoint. It accepts too that a state agency must exist to deal with these breaches of the peace, and that in order to do so, this agency must have licence to use asymmetrical force over others. In other words, a populace that has given its consent to be policed accepts, save for those rare and atypical moments when there is a genuine threat to the integrity of the underlying order, that in civil life, the police are uniquely entitled to use force, that when they arrive on the scene, everyone else relinquishes the entitlement to use force against them. As the mid-eighteenth-century French policeman Guillauté put it in his memoir of policing written three-and-a-half centuries ago: 'Citizens must be submitted to authority, disarmed internally and externally peaceful, without great alarms or pressing needs, before we may undertake to police them.'

To put it another way: a precondition of democratic

policing is that there is a demand for it among the general population. The police come to a scene because civilians have called them there, and they have called them because they want and need the presence of an agency that will use force or the promise of it to diffuse crises, to provide temporary solutions to immediate problems, or to subdue people who are menacing or dangerous to others. To obstruct a police officer from doing these things is, in all imaginable jurisdictions, a grave criminal offence that carries a prison sentence.

What happens to policing when it is no longer performed with the consent of those who are policed? The answer is fairly simple. In one way or another, the police retreat. They either avoid policing in those zones where they are not welcome, as Constables K and N tried their very best to do on the Saturday night I accompanied them. Or they use one or another means to negotiate their presence in those zones. To negotiate usually means to sell something: information, the obstruction of justice, the assurance that they will not intervene. For instance, it is quite possible that Constables K and N will, some time in the future, be tolerated in the vicinity of Toekomsrus's shebeens on condition that they obey certain rules: like never policing Darren and his friends, or policing their enemies especially hard.

The more police officers negotiate, of course, the more they begin to resemble other, private users of violence, and the less they look like police. While the uniforms, the two-way radios and other paraphernalia of policing remain, if the general population is not pacified, the idea of the police itself fades, and those who wear the uniforms and speak into the radios become both more and less than police officers.

The central argument of this book is that, 13 years after the inauguration of democracy, South Africa's general population has yet to give its consent to being policed. The instance of Darren and the two constables is but one example, and, on its own, not nearly strong enough to make my point. It could be argued that an example centring on the very young and the very delinquent in the early hours of a Sunday morning is not representative of much. In the pages that follow, I present layer upon layer of examples: I will show that the line separating police officers from the scores of private enforcers and protectors who litter South Africa is uncomfortably blurred.

Why is this so? Some of the reasons are, of course, historical. In the wake of the student uprisings of June 1976, the police were forced out of some urban townships. They were evicted from most of the rest a little less than a decade later during the insurrectionary period of the mid-1980s. Even before they were thrown out, the policing they provided was grossly inadequate: they did little to provide township residents with a bare modicum of personal security, and were in fact often among the various agents that periodically violated it. But with their forced removal from everyday life, the history of security in black, urban South Africa entered a new phase. In the vacuum left by the police, anti-apartheid politics and the politics of self-protection often became indistinguishable. All public structures and associations in township life provided protection, no matter what else they provided. And they all provided security to some people, never to everybody.

It was to this terrain – one where security was bought, sold and bartered, and also exchanged for solidarity and friendship – that the police returned in the early 1990s during the transition to democracy. My contention is that they never found sufficient moral authority to rise above the logic of this terrain, nor to refashion it, and that they thus had to negotiate their way into it and join its other players.

There are many reasons why this is so: some have to do with the strategic decisions South Africa's new rulers made about the police in the mid-1990s, others with who has been managing the organisation and how. But among the township residents I interviewed most gave two reasons. The first is that the police were never forgiven for their role under apartheid. They returned to the townships in the early 1990s a disgraced and ingratiating bunch, and never recovered their dignity, certainly never enough to become the agents by whom the general population would consent to being policed.

Second, police officers find themselves somewhere near the tail end of a frantic, unseemly dash to join the new black middle class. The middle class is a very expensive zone to occupy right now: what with pricey schools and homes in the suburbs, being middle class is far, far dearer than it has ever been before in South African history. The police are among a large category of township people who aspire very much to find a place in this class, who do not earn quite enough to get there, and who thus live beyond their means. They are, as a consequence, widely reputed in township life to be among a new breed of scavengers, prone to corruption and to the most expedient and instrumental attitudes to their own vocation. By virtue of their class position and their social aspirations, they are denuded of the authority required to do their work.

*

Where does poor Gladys, the chronic rent defaulter, fit into this history?

What happened to Gladys represents a very important exception to the story I have just told. For there is one sphere of urban life where a police presence is constantly demanded, and where the police are indeed conferred unbridled authority: in people's homes. Anyone who has spent any time on patrol in urban South Africa will know that the vast majority of calls to which the police respond are domestic, and that when the police walk into a private home, having been called there, their authority is immediate and absolute. They become, as Constables K and N did on that early morning in June 2004, blunt, authoritarian schoolteachers shunting their prepubescent charges around.

Gladys is an atypical example. In my experience, cops are called into private homes primarily by women whose partners are threatening or beating them or their children. But the command the cops exercised inside Gladys's flat is emblematic. They are serially called into people's homes and once there are given the unconditional authority usually reserved for adults over children.

That this is so is remarkable for a number of reasons. For one, it is new. It is something that has come with democracy. Cops under apartheid did not police everyday life. They certainly stormed into people's homes and took command there, but seldom to solve problems and seldom because they were called. So what the police do now, perhaps as much as 90 per cent of their time, they did hardly at all in the past. In that sense, the coming of democracy immediately turned policing inside out. There has, for generations, been a dammed-up demand for agents of order and authority to come into homes and defend the desperate; with democracy the dam wall burst.

It is remarkable too for its juxtaposition with what happens to the police out on the streets. In homes, their presence is demanded and they are in absolute control. By virtue of the uniform he is wearing, and by the fact that he has been called, a 25-year-old constable has the authority to treat elderly men and women with contempt. Outside the front door, the same constable is scorned and belittled, his uniform inciting hostility or indifference, the slivers of authority he possesses acquired through careful negotiation.

How does one explain this paradox? Perhaps it is a residue left over from generations of white rule: an old code in township DNA which remembers that the police can be challenged when people assemble in crowds, but are untouchable when they march into a home. In any event, in contrast to the past, they come into homes now because their presence is serially demanded by the vulnerable.

Indeed, the police and the vulnerable are natural allies. That they are so is counterintuitive and strange, for they also loathe one another. The vulnerable blame the police for treating them with little respect and with not nearly enough regard for their needs. The police, in turn, are prone to use their presence in private homes as an occasion to express disgust at the way people live their lives. And both sides, the vulnerable and the police, are acutely aware of the deadening emptiness at the heart of their relationship. Both know that the police do not possess the tools to bring lasting solutions to the problems they are called upon to attend; both know that the same conflicts will be recycled again and again. The vulnerable and the police thus form something of a reluctant family: they encounter one another over and over in the unhappiest of situations.

Nonetheless, it is towards the vulnerable that the police gravitate. You can see it during the course of a single shift. They seek the vulnerable out. They do so because the weak and the put-upon are a haven that shelters the police from the streets; because the world of the vulnerable is the one place where their presence is truly and unambivalently demanded; and because, in a difficult and complicated manner, it is through witnessing the pain of the vulnerable that the police have the opportunity to express themselves as moral beings.

*

I initially imagined this book forming two parts: one on urban cops, the other on the policing of rural life. As often happens, the project narrowed as it proceeded; an account of rural policing, I realised soon after I began writing, would have to wait for another time. And so, although not my initial intention, this has become a book about Johannesburg. We begin in Alexandra township, then reel back in time to sit in on a few key episodes in the policing of townships under apartheid. Next, we return to the West Rand, where Constables K and N work, then join a uniformed constable and a detective on the East Rand, before finally settling in an upper-middle-class suburb in northeastern Johannesburg.

The bluff

Another weekend night patrol, this one three years after Constables K and N's debacle, and about 45 kilometres northeast of Toekomsrus. I am in Alexandra township in northern Johannesburg, among the oldest black residential spaces on the Reef, and undoubtedly the most densely populated.

The police officers along with whom I am riding this evening are Inspector L and Sergeant Z. Both have been working in Alexandra for more than 15 years. They have before them a patrol plan. Hour by hour, it tells them what they should be doing from the moment they hit the streets at 7 pm until they knock off at seven the following morning. The plan was generated by Alexandra police station's Crime Information and Analysis Centre, which is equipped with some of the finest crime-mapping software available anywhere in the world. Each day, it imbibes a great batch of data detailing every crime reported in Alexandra in the last 24 hours, and every afternoon it spews out the station's patrol plans based on its analysis of the current distribution of crime.

The police can do wonderful things with the sort of software they have in Alexandra. One can, for instance, feed the computer a simple epidemiological description of all 89 murders committed in Alex over the preceding year, and the computer will produce an extraordinary analysis for you. It will tell you how murder is distributed between informal settlements, hostels and formal housing; between lit areas and unlit areas; indoors and out; before midnight and after midnight; in summer and in winter, around Christmas and during Easter; whether murders predominate in the vicinity of shebeens, and if so in shebeens where Zimbabweans drink or where South Africans drink; whether in neighbourhoods near the highway or near the river; in areas inhabited predominantly by old Alex families or by newcomers. If you know Alex well and you study the data for ten or fifteen minutes, you will have a pretty sound knowledge of the most important situational triggers of violent death in your jurisdiction.

Tonight the software has been used to shape Inspector L and Sergeant Z's patrol plan, and they treat it as sacrosanct. They pore over it for some time, make copies of it, file some of the copies, and slip others into their notebooks. They give me a copy, explain that it is a map of what we are going to be doing for the next 12 hours, and advise that I study it. It says that between 19h15 and 21h00, Inspector L and Sergeant Z will conduct vehicle patrols in Sectors One and Two of the Alex police jurisdiction; between 21h00 and 23h00 they will join other patrols to do a cordon-and-search operation at Ghanda Centre and the old Council Building, searching in particular for stolen goods, firearms and drugs; between 23h00 and 01h00, they will stop and search pedestrians for firearms and stolen goods between London and Rooth Roads; between 01h00 and 03h00 they will visit Millie's Tavern and Pat's Tavern on the

East Bank, as well as Alex Club and Capert House on First Avenue, to enforce the regulation that drinking establishments close at 01h00.

'Do you stick to this plan religiously?' I ask Inspector L.

'Not quite,' he replies. 'We get a lot of domestic violence calls on a Friday night, and we must respond to them. But when we are not responding to calls, we stick with the plan.'

We do not stick with the plan. Our movement over the course of the night follows a tight, easily discernable logic, but it has nothing to do with what the computer has generated. Inspector L and Sergeant Z know that very well from the start. Perhaps it is the expense of the software and the many people employed to service it, or perhaps it is because the station commander has decreed that this is how the software ought to be used: but the respect the two officers show towards their plan is as austere and as complete as its irrelevance to the way they spend their evening.

From the perspective of Inspector L and Sergeant Z's patrol, the two most important features of Alexandra are that on a Friday evening the streets are very, very crowded, and that most residents go to bed early.

Alex occupies some ten square kilometres of north-eastern Johannesburg. Nobody knows its precise population: there is enormous demand for space in Alex among the urban poor, and its numbers grow a little every day. The last census, conducted in 2001, put the figure at about 340 000. That's 34 square metres of ground space per resident: not a lot in which to move around.

The widest stretches of public space are the streets, and it is the streets people use whenever an event or a ceremony brings together a crowd. Funeral tents are erected across public roads, and so the death of a resident turns thoroughfares into cul-de-sacs, diverting the township's traffic. Partygoers often block off a section of the street with bricks and broken bottles, assemble their speakers on the edge of the pavement, and turn a ten-metre stretch of road into a dance floor.

If the computer had generated a plan that corresponded to the actual imperatives of Inspector L and Sergeant Z's patrol, it would have begun thus: 19h00-00h00, avoid crowds. In Toekomsrus, that meant skirting two well-known landmarks. In Alex on a Friday night, it is not that easy. One cannot anticipate where the crowds are, and one is given little warning before one finds oneself in the thick of one. Driving through a shack settlement of Mozambican immigrants at about 8 pm, we turn a corner to find that we are facing a group of 100-odd people. They are standing in wide, lopsided circles in the road, most of them clutching bottles of beer. From the speakers that they have mounted on a nearby roof, bass-heavy house music slams into the tin walls of the shacks and rebounds back into the street. We slow down, edge gently into the front of the crowd, then stop.

The police officers' dilemma is this: drinking on the street is illegal, and, by rights, those holding bottles of beer should be arrested. In South Africa's townships, enforcing the law against drinking in public is deemed very important. Alcohol consumption breeds a raft of different crimes, from street fighting, to mugging, to violent domestic disputes. Both the perpetrators and victims of murder are in most cases under the influence of alcohol. Restricting the time and place of its consumption is the very spine of weekend crime prevention.

Yet here and now, trying to make an arrest isn't an advisable course of action for Inspector L and Sergeant Z. The crowd would never permit any of their number to be taken from them and thrown into the back of a police van. And yet driving on, heads down, as if the people around are not breaking the law: that is so transparent a display of their impotence that Inspector L and Sergeant Z will not countenance it. And so each gets out of the vehicle, finds a man holding a beer bottle, gently wrests it from his hands, pours its contents into the street, and puts the empty bottle down at the side of the road. The two men who have been chosen for this treatment watch impassively. Those around them fall silent and stare with cold and expressionless eyes. Most sip every now and again at their own beers. Inspector L and Sergeant Z appear to know better than to try the same with anyone else. Two lost beers are as much as this crowd will tolerate. The officers make their way back to their patrol van and drive off.

A few minutes later, the Mozambicans now some distance behind us, Inspector L catches my eye in his rear-view mirror. In the moment our eyes lock, he finds the humiliation he is feeling in my gaze.

'There are many of them,' he volunteers unhappily. 'You try to take them in for a small crime like drinking, and you create a lot of trouble. A lot of trouble for such a small crime.'

That much is certain. Less clear is what has just happened. The officers did what they did to salvage some dignity. The crowd humoured them; they were prepared to watch two bottles of beer seep into the ground so that the police officers could go through the motions of saving face. Why was the crowd prepared to do that? A simple cost-benefit calculation, no doubt. If you beat up two cops, the police must retaliate. The following evening they will throw a cordon around the shacks and descend upon the Mozambicans in large numbers, backed by air support and a phalanx of vans waiting to be filled with arrestees. Two lost bottles of beer is not worth that.

But why not? If the police know that pouring two bottles of beer into the street comes at the cost of two injured officers and an airborne operation, they will surely stop emptying beer into the streets. Of the two sides to this relationship, the police and the Mozambican community, it is the Mozambicans who ultimately decide to what extent they are policed. Here and now, they will lose two bottles of beer to the street, but not two people to the police station cells, in order to assist to recoup some of Inspector L's dwindling self-respect. That is today's threshold; no doubt it wobbles and shifts: the unwritten rules are drafted and redrafted.

*

About an hour later, I watch a re-enactment of the same drama: Inspector L and Sergeant Z pretending to police, the young men of Alex pretending to be policed. We are in the heart of old Alex, driving away from a complaint.

About us, the people on the pavements are moving, some briskly with heads down, others at leisure. Nobody is alone. Some walk in pairs, others in groups as large as half a dozen.

Three young men are standing in a circle on the pavement. As we pass, one of them slips a knife into the back of his trousers. It is poor timing: had he done that a moment sooner or later, we would not have witnessed it. But we have, and the three youngsters immediately see that we have. They turn on their heels. Sergeant Z reverses, spins and chases. He does so with remarkable agility, and we are soon gaining ground on them. Two of the boys peel off into alleys, leaving the one with the knife in our headlights. He is in a cul-de-sac now, our brights on his back, the road in front of him at an end.

And then Sergeant Z and Inspector L do the strangest thing. Instead of using the bulk of his van to corner his prey in the cul-de-sac, Sergeant Z stops some ten metres away. The boy's eyes dart around, searching for the escape route Sergeant Z has offered him. Both cops reach for the guns in their holsters, then open their doors, but their movements, while as fluid as ever, are now languid, as if they have suddenly entered a parallel world in which things are done a little slower. They reach the street in time to see the youngster's white sneakers disappear over a corrugated iron wall as he flops, headfirst, into an old Alex yard. Sergeant Z, the younger and thinner of the two police officers, scales the wall too, but it is just for good measure, and he returns a few moments later and shrugs.

As we get back into the car, I look around and see that a good 30 or 40 pairs of eyes are watching us. When we turned around and gave chase, everyone on the street interrupted their journeys to await the outcome. To my left, two teenaged girls are chewing hard on their gum. Behind them, four men and a woman are huddled around a fire. They are all staring at us. What in one sense was a pursuit of an armed man was in another a performance before a large audience.

Dimly, or perhaps even vividly at the front of their minds, Inspector L and Sergeant Z knew that the audience was partisan, that it was not on their side, and that should they win the crowd may not tolerate the outcome. Indeed, they must surely have known from the moment they spotted the knife that this street was too crowded to allow for an arrest, that the boy with the knife was never going to end up in the back of the van. Nonetheless, he saw that they had seen his weapon, and so the game was on, they had to act, because that is what cops do, because to have him see you see him with a weapon and do nothing about it is just too shameful.

Perhaps it is more than shameful, and perhaps that is indeed the secret of both of tonight's events. I am speculating freely here, but maybe it is this: For a cop to see and yet to do nothing is to cross a threshold beyond which the sheer nakedness of his impotence becomes intolerable to those around him, and it is no longer safe for him to be on the streets. And so, while Alex is crowded, the cops must bluff, and the crowds must not call the bluff.

I said earlier that a precondition for policing is the consent of the general population to be policed. In this sense, the cops are always bluffing, and the role of civilians is always to refrain from calling the bluff. It is the citizenry who determine to what extent they are policed.

At this time in South African history, the Friday-night populace of Alexandra permits a two-officer patrol only the pretence of policing, and no more. Inspector L and Sergeant Z have learned to walk the narrowest and most delicate of ropes. If they see a knife in the crowd and apprehend the culprit, they believe that they will be overpowered. Yet if they are seen to see a knife and do nothing, their presence will not be tolerated either. And so they are actors in a theatre whose script they have had little say in shaping. But they must know their lines very well.

Why do the crowds play along? Why is this the script that has been written? Perhaps because to call the bluff and reveal the cops in their true nakedness is to recall a time to which few people want to return. It was not so long ago that Alex crowds refused to allow two police officers to drive freely around the township on a Friday night. That memory is fresh in everyone's minds.

Perhaps what the crowds are saying is that they do very much want to be policed, but not by this police force.

*

Both of these events take place between darkness and midnight, when the volume of people on the streets is at its highest. Occasionally, during this time, a complaint of domestic violence takes us away from the crowds and into a private home.

A remarkable feature of a Friday-night Alex crowd is its shape. Along the wide avenues it is fulsome and thick, like the throbbing body of a fleshy beast. But it is not confined to the avenues: it leaks off into yards and alleyways and footpaths, and one gets the sense that were one to follow any of these thin tentacles, one would discover that the crowd has no end; that following a frail limb or antenna of crowd as it turns and weaves, one would eventually link up with another writhing avenue; that the entire crowd, perhaps a hundred thousand strong, is one intricate network with no outlet and no end.

From this labyrinth of crowd there is but one escape: the indoors. You walk into somebody's home, and when the door shuts behind you, the crowd is gone: the windows are covered with cloth and curtain, and so you cannot see it; the house is sealed against the cold of winter, so you cannot hear it. But you know that it is there, and your awareness of it turns the place you have entered into an underground den or a womb: a place of exquisite shelter.

And that, I think, is in part what a complaint of domestic violence on a busy night means to the police. They are on official business here, business they are obliged by law to attend to, and they thus have legitimate reason to close the door on that menacing crowd. For the next half an hour or so, they need not be humiliated and scorned, nor have eyes in the backs of their heads. The three domestic violence calls Inspector L and Sergeant Z respond to between darkness and midnight are, I think, a reprieve. In these violence-wracked dens, they are here because they have been called. They are thus in full control; it is their authority, after all, that has been demanded. It is for them to decide how the crisis they have walked in upon will be resolved, and they are buoyed by the strength that has been conferred upon them. There is violence here too, but it is almost always that of one man, whom they outnumber, and he is easily subdued.

There is a natural, albeit deeply unhappy, affinity between police officers and vulnerable women. They inevitably gravitate towards each other: the women because they have lost control to violent men and are in desperate need of a greater strength to stand at their side; the police because the world of abused women is their refuge. The two groups resent and seek one another in equal measure.

*

The third domestic violence complaint we attend to takes as much as an hour, and by the time we are back on the streets, they have begun to empty. It is about 15 minutes past midnight. By one o'clock there is barely a soul about, even on the main avenues, which, just 90 minutes or so earlier, were full. I had not taken Alex for a place that beds down early.

These are by far the most dangerous hours of the night. If the crowds spell menace for the police, they promise protection for civilians: it is very rare in Alex for a person to be mugged under the gaze of a crowd. It is after midnight that stragglers are held up at gunpoint and robbed, and women raped. On my second Friday night in Alex there were, unusually, three murders, and they all took place in public spaces between one and two-thirty in the morning.

Inspector L and Sergeant Z are visibly more at ease. For the first time tonight, one of them finds a CD and plays it, and for the rest of the shift we listen to an eclectic medley of South African male vocalists: Vusi Mahlasela, Johnny Clegg, Sipho 'Hotsticks' Mabuse.

With the crowds gone, and the old men of South African music to gird them, the two officers start to take their revenge on Alex. The engine growls as Sergeant Z picks up pace, and the cops begin, for the first time on this shift, to do what their patrol plan has suggested they do: stop and search. They choose only men walking in pairs, not larger groups. They pick them out in the headlights at a distance, then move in very fast, swerving in close, cutting right in front of them and slamming on breaks. Before the car is stationary, Inspector L has jumped from the passenger seat and has one of the young men by the collar. If they have not seen us coming, there is no time to escape.

But even now, with the streets near empty, Inspector L and Sergeant Z exercise extreme caution, and the choices they make are revealing. At about 2 am, we are cruising down Vasco de Gama Street, which divides Alex from the industrial district of Marlboro. Walking in the same direction some distance ahead of us are three young men, and, some twenty paces in front them, three young women. The young men wear baggy trousers and big jackets, and each is swaggering down the street with the hyperbole of those who need to show ownership. In contrast, the three women, who seem to be high school girls, have linked arms, are singing very loudly and quite beautifully, and are moving in step.

As we reach the young men, Sergeant Z swerves close to them and slows down, and Inspector L leans out of the window to look at them closely. They know that we are there, but they have chosen to pretend not to notice, and the two police officers read this brazenness as trouble. If they try to stop and search these men, they calculate, they are sure to resist. And perhaps they do indeed have a gun. Outnumbered three to two, the officers are not comfortable. So we drive on, and moments later we are alongside the singing girls. Inspector L rolls down his window and begins to address them in a coarse, irritable voice.

'Go home!' he hisses in Tswana. 'Get off the streets! You are going to get yourselves raped.'

Two of the three girls glance momentarily at Inspector L and then look ahead again. Their singing does not pause. Sergeant Z drives on.

This is by no means unusual: a South African cop on a township street scolding young women for being out too late. In Kagiso township on the West Rand a year or so earlier, the police officers I was with insisted on giving every young woman they found on the streets after midnight a lift home. But the juxtaposition between this scolding and the tactical withdrawal from the three young men is striking. It seems to me an analogue of South African patrols in general. Between the perpetrators and victims of crime, it is engaging with the former that carries the risk, so it is better to engage with the latter. And thus, for thousands of township women, their recurring experience of the police is of ill-humoured, inarticulate nannies in uniform shooing all and sundry indoors.

*

At about 1 am, we come across a scene best described as lovely. Leaning against a wall on the side of the road are two young lovers. The man is wearing an outlandishly large coat. The woman has wrapped herself around his body, and he has wrapped his coat around her body, and they are pressed together and largely invisible inside the great tent that they have made.

'Get off the street!' Inspector L spits from his place in the passenger seat. 'Go!'

The man's head emerges from the mass of coat and flesh. He is smiling broadly and charmingly, as if to say: 'Have a heart.'

'Fuck off home!' Inspector L hisses, this time more threateningly.

The lovers do not disentangle, but begin to retreat indoors as one, her feet resting on his, his legs moving in slow, lumbering steps, like a man in a space suit, as he carries the combined weight of them both.

Their gentle, comical air makes of Inspector L's snarling face something so ugly that I smile to myself. For the second time tonight, he is watching me in his rear-view mirror.

'They are a danger to themselves,' he snaps. 'Four tsotsis will come and want to rape that woman, and what will that young man do all on his own?'

We drive in silence for a while.

'And anyway,' he continues, 'what they are doing is

not right. It is disrespectful to the community. What is happening under that coat is more than kissing, and it is disgusting to do that out on the street.'

He is about to remonstrate some more, but then thinks better of it.

It strikes me that what Inspector L has just done is quite complicated. He cannot police the township's predators because he has neither the strength nor the authority. What he can do is police the weak and the harmless: battered women who call him to their homes, young women returning from a night out, two lovers on the street. He does not like the fact that he has been reduced to this, and he expresses his dissatisfaction in the form of a grumpy, intemperate lashing out. He has become not so much a policeman as a conduit for the expression of moral disgust. He is like the crank who habitually calls in to talk-radio to express his displeasure with the world.

Captain R

I return to Alex the following night. This time, the patrol I am to accompany is 15 officers strong. At 8 pm, Saturday's busiest time, they leave the station together and walk down Fourteenth Avenue in a jagged line. In these numbers, they own the space through which they move. They stop and search at will: young men on foot and in vehicles, drinkers in street-side shebeens. There is not a trace of the fear Inspector L and Sergeant Z exuded all of last night. The shoe is on the other foot now.

And so I learn a lesson that any South African cop knows: where a population is reluctant to give its consent to being policed, you police it by outnumbering it.

The patrol's commander is a great bear of a man called Captain R. He has been posted at this station since 1980 and is well known in all the streets we are to patrol. Many of the men arrested tonight will plead with him, addressing him respectfully as Bra R. He will not even deign to make eye contact with them.

Captain R trails his men and women in a white Volkswagen CitiGolf, his headlights catching the metallic twinkle on their epaulettes as they move through the crowd. When somebody is arrested, he is put in the back of Captain R's car. When the car is full, the entire crew moves back to the police station to book the arrestees into the cells. Then they move out into the township again.

With this new configuration of numbers between cops and civilians comes a new set of rules. Nobody who is stopped and searched resists or displays any displeasure. He raises his arms with quiet compliance as he is patted down, and stares ahead. Sometimes, he casually continues a conversation with a companion that the police have interrupted. Once the cops are done, he nods and moves on.

But when the police decide to arrest, it is a different story. Nobody who is arrested goes willingly. Each shouts volubly at the top of his voice, throwing his fists and his feet at every advancing uniform, sometimes using his forehead and his nose.

It is a pointless exercise: he is outnumbered fifteen to one, nobody in the crowds will step in to help, and all that can happen is that he will come to grief. And if the officers are taking too long to subdue him, Captain R alights from his Volkswagen CitiGolf and lends his considerable bulk to the campaign. But resist the arrested one must: it is an ethical imperative, it seems, a necessary performance before one's peers in the teeming crowd. But even after the man has been subdued and is in the vehicle, and the crowd is no more, he continues to remonstrate volubly.

'Bra R,' one of them drones lugubriously, 'I swear to God, this place is going the way of Zimbabwe. I swear it. To be bundled off the street without any reason, we are going one way, the way of Zimbabwe.'

The man rattles on, repeating the same complaint. Bra R ignores him for a long time. When he can take it no more, he turns in his seat, puffs out his chest, and hurls at the drunken man a voice so thunderous that the windows tremor: 'SHUT UP, FOR GOD'S SAKE!' he roars, and that is the last we hear of Zimbabwe.

Everyone jailed this evening has been arrested for public drunkenness or drinking in public. The prize arrest would be a knife or an illegal gun, but tonight there are none. Which may be a sign of a growing knowledge that it is no longer a sure bet to walk the Saturday night streets of this township armed.

*

After midnight, when the streets are emptying, Captain R and his crew take a different tack. The foot patrols cease, and the 15 move through the township crammed into three cars, Captain R's at the front, the other two forming the legs of a stumpy, inverted V on either side of him.

Captain R decides whom to stop and search, and he does so by aiming his car straight at them. The others follow, closing their prey into the V, and the officers jump out as their cars screech to a halt. In the face of such aggression, their targets invariably flee, that they have done nothing wrong and have nothing to hide notwithstanding. Captain R's charges sprint after them into the night.

After an hour of this, the foreheads and cheeks of the fit young constables under Captain R's command are glistening with sweat, their diaphragms hungrily sucking in great gulps of air.

The vitriol and testosterone this heavy policing brings

to the street emits a bad vibe, and as the night grows older, the people we encounter are drunker and more intemperate. By 1 am, each encounter is sour and sharp-edged, and by 2 am, the officers have decided to move into groups with guns drawn. There is something foul and unpleasant in the air.

An unusual call comes over the radio that is to draw Captain R's team away from this circle of ill feeling that they have spun. A man has walked into the Alexandra police station saying that a drinking companion took out a gun earlier in the evening and threatened to kill him. He said he did not believe that the drinking companion had a licence for his gun: the drinking companion bragged of having just bought it, that day, on the black market. The complainant left the man's name and address at the police station.

Led by Captain R, all 15 officers climb into their vehicles and make their way to the address the radio dispatcher has given. We walk into an old Alex yard, and the tenants in the shacks and rooms at its edges come out to watch. Two officers find places for themselves deep in shadow and draw their guns. Four more cluster around the front door. One of them knocks. The house is in darkness. Nobody replies.

Among the tenants in the audience, someone volunteers that the people in the house we have surrounded left for a shebeen more than an hour ago. She points to a maze of shacks on the other side of the street. The officers leave the yard, disappear into the shack settlement, and emerge some ten or fifteen minutes later empty-handed.

Captain R, who has spent this time waiting impatiently in his car, sighs and looks at his watch and shakes his head. The shift is almost over, and this call has taken his men and women off the streets for too long and without reward. We head back to the station for the last time.

I am to discover later that this case is not followed up. The name of the alleged owner of the illegal weapon is not recorded, and the address we have visited is to vanish into a pile of forgotten logs. One would have thought that the name and address of a man who may just have bought a gun on Alex's black market is a gold mine, that it is worth a considerable investment of the station's resources and time.

And so it strikes me that the purpose of tonight's patrol is not transparent: not to me, not to the police, and not to the residents of Alex. Yes, it is certainly about keeping weapons and alcohol off the Saturday-night streets and, in that sense, it is about reducing violent crime. But the horizons of the patrol's purpose are no longer than the duration of the patrol itself; it is as if, once the sun has risen and tonight's officers are sleeping in their beds, that illegal weapon no longer exists. And so perhaps tonight's activities are animated by manifold strands of purpose. One is the simple display of aggression, the exercise of dominance, the satisfaction of gathering in numbers in a place where patrolling in twos leaves police officers humiliated and vulnerable.

Tonight's patrol is, in part, an exercise in revenge.

*

With his officers in the police station booking in the last arrestees of the night, Captain R and I drive through the streets alone. Detached from his army, the adrenaline still pumping through his great body, he uses his accelerator as a substitute. He cannot stop and search people now, but his tyres can still squeal. As we race down the avenues, the streets' remaining pedestrians stop and frown.

I ask Captain R where he went to live when the police officers and their families were chased out of Alex in the mid-1980s. He turns briefly to me and raises an eyebrow, then stares straight ahead.

'They made temporary homes for us at Leeukop Prison,' he replies. 'My family ended up staying there a number of years.'

'When did you move back to the township?'

'I never moved back. I bought a house in the suburbs on the East Rand.'

We drive along in silence for some time.

'How long has it been possible,' I ask, 'to do the sort of patrol you did tonight?'

He thinks a while. '1999, I'd say. Before that, if you tried to do what we did tonight, you would get shot.'

To Newclare and back

And so this weekend with Captain R, Inspector L and Sergeant Z in Alex in July 2007 is a moment in a history between residents and police, and a fast-moving history at that. One way to begin to reflect on it is through the story of the prosaically named Sector Four Patrol Group.

Every Friday and Saturday night at nine o'clock, between 50 and 60 men gather at a house in Twelfth Avenue. They are almost all elderly or middle-aged. If you know the township well, something else about them immediately draws your attention. All are of old Alex stock; they are members of families that have lived in the Alexandra yards for generations. And if you know Alex particularly well, you will note that the vast majority of these men were activists in the Alexandra Civic Association and the Alexandra Youth Congress in the 1980s; they are veterans of the Six-Day War of February 1986, when the residents of this township rose up against the apartheid authorities, threw them out, and briefly governed the township themselves.

By 9.30 pm, the men have divided into three or four equal-sized groups, and have headed off in different directions. At least one person in each group has a licensed firearm tucked into his belt, and a cellphone to keep in contact with the other groups. They move through the crowded streets and the shebeens, and sometimes through the shackland labyrinths, stopping and searching men for weapons. It is by anyone's standards very dangerous work; if a culprit is not quickly disarmed he will use his gun to free himself. The patrol began its work in 1999, and in the eight years since, three of them have been killed, and one paralysed, while many others have suffered minor gunshot wounds. They have, since their formation, taken 610 illegal firearms off the streets.

Watching them move through the Friday-night crowds, there is a plodding steadiness to their rhythm, an aura of elderly men at work. Strictly speaking, what they do is illegal; civilians do not have the authority to stop people in public spaces and demand to search them. And yet that they are long-time residents rather than police officers gives their relation to those they search a quiet evenness. When they walk into a shebeen in a lumbering group of 12, and search every drinker in the room, the air does not seem to crackle with rancour as it does with Captain R; the drinkers submit to the hands patting down their bodies with absent-minded indifference before settling down again to their beers.

*

At lunchtime in a small square room at the old magistrate's court building on the border between Alex and Wynberg, I interview Sector Four Patrol Group's chairperson, Dan Sibanda, and its media officer, Joe Mukwali. I want them to speak about the connection between what they do now and what they did 20 years ago; what is the thread, I ask, that connects an uprising against apartheid then with ridding the township of illegal guns now.

Bra Dan is amused by the question. 'We are the ones who threw the police out of this township,' he chuckles, 'and now we are the ones who have invited them back. It is only appropriate. And I do not exaggerate when I say it is us who invited them back; before we were formed in 1999, it was not possible for them to walk down a street in this township. It is we who have made the streets safe enough for them to patrol.'

Bra Dan is, I would imagine, in his mid- to late sixties. He speaks slowly and with exaggerated care, pausing often to think. He has come to our meeting in an ancient Mercedes Benz, a great many years of repair work under its long, square bonnet. Those who have formed the Patrol Group are of a particular ilk. They are working class and not especially well educated, and a sizeable number of them are unemployed. They are not among the struggle's beneficiaries: not among the young, ambitious Alexandran activists who have subsequently left the township for the suburbs and gone into business or government. Their patrol, which costs several hundred rand a week to maintain, is financed from their own shallow pockets.

'The transition years were a funny time,' Bra Dan continues. 'The war was over, we were putting our energy into supporting the ANC at the negotiating table, and things relaxed a bit. There was no longer a position among the civic and the youth structures that the police must not be here. But still, they could not move freely; if they walked in the township they would be attacked, and not just by the criminals, but by ordinary residents. And it was very, very rare for people to call them when there was trouble.'

I interviewed many other Alex residents about their memories of the police during the period between the unbanning of the ANC and the release of Nelson Mandela in 1990, and the first democratic election of April 1994. One of them, Joel Sebolao, was in his late teens and early twenties during those years. He recalls the local officers from the Alexandra police station as meek, scared, and utterly lacking in moral authority. The cops township residents feared, he says, were the detectives from stations in nearby white suburbs who would make sorties into Alexandra looking for the perpetrators of predatory crimes.

'Brixton detectives came in to look for vehicle thieves,' Joel remembers, 'Bramley, Sandringham and Norwood detectives for housebreakers, and the security police from John Vorster Square came looking for comrades. And they were brutal. If they thought maybe your brother-in-law was the perpetrator of the vehicle theft they were investigating, they could knock on your house at three in the morning and beat you until you told them what they wanted to know.'

Those who policed the white suburbs came in as foreign invaders; those who policed Alex were too afraid to walk the streets of their own precinct.

I ask Bra Dan to pick up his chronology where he left off:

'It was only after April 1994,' he continues, 'that the first real contact with the police began. The station's first black commander called a meeting with community structures and said he wanted to build a partnership with us to fight crime.

'A general meeting was called at Alex Stadium. Thousands of people attended. The question of a working relationship with the police was discussed. Every single detective from the station was there. The executive of the new Community Policing Forum was there. People said at that meeting that the number one issue in this township is crime. Most people said it was because of unemployment and because of the taverns. The police said they understand, but they are scared to police this place. They asked the community to assist them. We made lots of plans and met with them countless times. But another five or six years would pass before it was possible for them to walk on foot in this township.'

I asked a detective by the name of Inspector B, who had worked at Alexandra station since 1990, what changed in the wake of the big meeting at Alex Stadium.

'Members of the Community Police Forum moved right into the station,' he said, 'right into the charge office. And they watched everything we did. If somebody walked into the charge office to lay a complaint, a CPF member was appointed to see the whole case through, from beginning to end, to watch every step we took. That was the only way any resident would ever deal with the police. They would say: "Where is the CPF representative? I am not going to be alone in a room with you."'

The way Dan Sibanda and Joe Mukwali describe it, the patrol group formed because, by 1999, the prospect of the police ever having the freedom to control crime in Alexandra remained remote.

'The whole idea was conceived in the station commissioner's office,' Joe Mukwali says. 'We realised in that meeting with him that we in the civic structures and the police needed one another. We could protect them, and they could protect us.

'They needed us because the residents would not allow them to patrol. And we needed them because when we went up to people to search, they would say: "Who are you? What gives you the right to touch me?" And so we realised that we must patrol together, protecting one another.'

When the Sector Four Patrol Group officially launched in 1999 – 'Sector Four' refers to a sub-jurisdictional policing sector in Alexandra, a massive square of several thousand homes in the middle of the oldest part of the township – then Safety and Security Minister Steve Tshwete made a speech and joined the first patrol, a mark of its members' impeccable political pedigree.

'The residents were not yet used to us searching them,' Bra Joe says. 'We went into a shebeen to search and the drinkers fought back. The Minister was punched in the face. They did not know it was the Minister, just that he was searching them, and they hit him. He left immediately afterwards. It was about 11.30 pm at that stage, but we kept on doing our work until 3 am the following morning.'

*

When they tell it this way, Dan Sibanda and Joe Mukwali's tale is one of gradual progression from a time when the community and the police were at war to a state of normality. That the outline of the normal is beginning to take shape only now, during the ANC's second term of office, speaks to the difficulty of the transition, not to the doubtfulness of the journey itself.

In this scheme of things, it is appropriate that the ones who kicked the police out are those now bringing them back. Heirs of the township's oldest families – bearers of wisdom that comes only with age, and of authority that comes with their role in the uprisings, dependable because they are not among those who have flitted off into the middle classes and suburbs – they are the backbone of this community, its very substance. Of course it is they who are bringing the police back; of course it is they who are stitching together a normal relationship between a township and a democratic state.

There is some evidence to support this optimistic version of things. Some categories of violent crime have plummeted in Alex since the mid-1990s. Take murder, for instance: 241 reported cases in Alex in the 1995/96 financial year; 89 in 2006/07. Or recorded cases of assault with intent to do grievous bodily harm: 1 095 cases in 1995/6 down to 885 cases 11 years later. These are crimes committed primarily on weekends involving guns and alcohol respectively, sometimes both. That Alex is now a place where drinking is partially regulated and where one is no longer as free to carry a weapon through public spaces is surely the foundation of the declining rate of these crimes.

Other things support this progressive story. The old Alexandra police station, positioned in an industrial suburb beyond the periphery of Alex like an outpost of the white suburbs that lie to the west and the south, is now no more than a satellite building. The new station is deep in the very heart of Alex, wedged between the taxi rank and the high school that was once the epicentre of the uprisings.

And since Alexandra is the site of a presidential lead project, its police station is now among the best resourced in the country. The list of personnel and services at its disposal is truly impressive by South African standards. Whereas some township police stations have a detective service so small they have to call in outside units to deal with serious crimes, each of Alex's six subsectors has a team of ten or so detectives dedicated to it alone. There is a special detective unit dealing with robbery and housebreaking, and another dealing with domestic violence and rape. An NGO specialising in conflict resolution operates through the magistrate's court offering mediation in lieu of criminal prosecution. The station has a supplementary unit dedicated to tracing outstanding suspects in old cases. Each sub-sector has a fulltime manager whose sole job is to liaise with community members. There are at least six vans patrolling Alex at any one time. And there is a crime prevention intervention team – the sort I accompanied with Captain R – which works flexitime and hits the township streets when the number-crunchers and their fine computer software ascertain that they are most needed.

By this reckoning, the Alexandra of the apartheid

years no longer casts its long shadow over the Alexandra of the present. The township is now truly a different place.

*

And yet for all that, one need only begin to ask Dan Sibanda and Joe Mukwali a slightly different order of questions, and another story begins to tell itself. When I first met Bra Dan, we spent the day riding around Alex together in a patrol van with Inspector L, whose Friday night patrol I documented in the second chapter of this book. At some point during the day, Bra Dan said pointedly that he preferred white to black cops. I'd asked him why.

'They are straighter,' he replied. 'They do not swerve this way and that like black policemen do.'

It was a lavishly provocative thing for him to say, no doubt for Inspector L's ears. But when I tried to press him, he changed the subject.

When I meet him at the magistrate's court a couple of weeks later, together with Joe Mukwali, they both speak far more freely. The Sector Four Patrol Group claims to have taken 610 guns off the streets since 1999. Each time they find an illegal firearm, they perform a citizen's arrest, call the police, go to the station, and write a witness statement. There is little question that their claim of 610 cases is correct; in their files, each case is accompanied by the paperwork the ensuing arrest has generated.

And yet, when I consult Alexandra's official crime statistics, it emerges that between 1999 and 2006, exactly 900 people were arrested for illegal possession of firearms and ammunition, which means that more than two-thirds of the illegal weapons taken off the streets are taken by Bra Dan's people, rather than by cops. When I tell them I find this close to unbelievable, I am greeted by an uncomfortable silence.

'Well,' Bra Joe finally says, 'I'm not sure how accurate those statistics are. Some funny things have happened to some of our cases.'

'Like what?'

'Some cases disappear. We do the witness statements, there is a bail hearing, and then nothing. When we follow up with the station commissioner, he checks the record and it says: "case undetected". We say: "No. We brought you a gun *and* a suspect. We took him to you with our own hands." And then we pay from our own pockets the taxi fare to the records office in Gallo Manor, and it says there on the computer that Dan and I have been withdrawing cases. Not just one case: many cases.'

'And what else?' I ask.

'They have so many tricks,' Bra Dan replies. 'If the witness lives at 25 Eighth Avenue, somebody will add a number to the address in the subpoena so that it says 125 Eighth Avenue. And so it is written on the docket: "Case dismissed: witness untraceable."'

'Or,' Bra Joe adds, 'the witness will get a subpoena that says he must come to court on 12 February. He goes on that day to find that the case was dismissed ten days ago: the trial was on 2 February.'

'I don't understand,' I say. 'By now you must know that police station so well. You must know exactly who are good detectives and who are corrupt detectives. Surely you can assist in isolating the corrupt ones.'

'We do not know who they are,' Bra Joe replies. 'We open a case on a Saturday, it is handed to the detectives on a Monday, and we do not know whose case it is, or what they are doing with the case. Sometimes we only know the outcome because we see the suspect wandering freely around the township.'

'Surely you go to the station commissioner?'

'Yes, sometimes. And he says he will do something about it, and then nothing happens, and you don't want to complain about the detectives again and again because it creates estrangement, and the police are our allies. We must be able to work with them.'

'So you cannot say who in the detective service is corrupt? From your vantage point, when the docket goes to the detective service, it is being thrown into a black hole?

'There is a market,' Bra Dan replies. 'And we are invited into that market. Take an example. Last year, we caught a man with an illegal weapon on a Saturday night, and he tried to shoot us. His gun jammed on the third shot. We took him to the station, and opened charges of possession of an illegal weapon and attempted murder. On the Monday morning, his wife came to us and offered R3 000 to drop charges. We said no.

'A week later, we see him on the streets. We go to the court to see what has happened. He is out on R1 000 bail. And there is no attempted murder charge. That charge has just …' Bra Dan finishes his sentence with a click of his fingers; vanished into thin air, his gesture says. 'We can only assume that his wife offered the R3 000 to the investigating officer, or to somebody.'

Bra Joe now takes the reins. 'Another story,' he says, 'this one similar, but worse. We arrest a man for illegal possession of a firearm. This one's wife also comes to us. She is offering us R10 000 to withdraw charges. We say no. Two days later, the man is meant to be in jail, his bail hearing has not come up yet, and we spot him in the passenger seat of a car driving through Alexandra. In the driver's seat is the investigating officer of this very case. We phone the investigating officer immediately on his cellphone.

'"What's going on?" we ask him. "The suspect is meant to be in jail, but we have just seen him in your car."

'He says: "This man is my friend. Just because I'm the investigating officer, does that mean I must suddenly pretend that I have not been his friend for years?"

'Some time later, this fellow, the one who was in the passenger seat of the detective's car, is spotted driving through our area in a microbus with a big group of people. They are looking very threatening, very menacing. I take out my gun and point it at him and ask him what he wants. He says: "I am looking for Dan. Where is Dan?" I say: "What do you want with Dan?" He says: "No, I am just looking for him."'

'With 610 arrests under your belt,' I say, 'you must have a lot of enemies in this township.'

'Our biggest expense,' Bra Dan says, 'is to protect the people who are making statements. You see, when we make an arrest, the one who found the gun and made the arrest is the one who must write a witness statement and then appear in court. The friends of the one who has been arrested arrive at the court in numbers, and they threaten the witness. So we must also go in numbers. We must get transport to the high court in Johannesburg, to the high court in Pretoria. We must buy food for the day. Most members are unemployed, but it must come out of their own pockets. And it is necessary work. If we do not give moral support to the members who are under fire, nobody will be prepared to patrol anymore.'

'Already our numbers are dwindling,' Bra Joe says. 'The expenses keep mounting. Three weeks ago, a patrolman was shot in the foot. His foot is in plaster, he is not working, and we are still feeding him and his family. He is our burden.'

'So why do you continue?' I ask. 'What keeps the organisation going?'

'If we withdraw from the fight against crime,' Bra Dan replies, 'we will be in danger. If we stop, criminals will target us; there are people we have arrested several times. As things stand, I can leave my home for a month and go to the rural areas and nobody will come near it because I am Sector Four, because I am Sanco [South African National Civic Organisation]. If I leave the patrol group, I am very vulnerable.'

*

Dan Sibanda and Joe Mukwali are telling two stories, not one, and each vies with the other for dominance. In the first, the police are representatives of a new democratic state, their task to keep the peace among a citizenry that is slowly submitting to its own pacification. In this story, the Sector Four Patrol Group is an embodiment of that citizenry, a gathering of the wise and the elderly who take the police by the hand and help to install them in their rightful place as protectors of anyone and everyone.

In the second story, there are no police, not in the most meaningful sense, since the general population has not given up its right to wield violence, and what distinguishes the police from others is thus not sufficiently clear. Where there are no police, there are, instead, groups of people who bear weapons to protect themselves, others who are armed in order to sell protection to third parties, and still others who use their arms to police the markets in which they make a living. In this world, those who wear the uniforms of the state are not the police, but bearers of influence, information and power that is bought and sold on a market. And Bra Dan and Bra Joe are not the embodiment of a community's interests, but one among many groups that has come together bearing arms to protect itself from the hostility of others. This is certainly not a judgment cast upon them; they do not choose in which of the two stories they are to appear; that is chosen for them by the other players.

Neither story eclipses the truth of the other. They jostle side by side, each one real and alive. The scary thing for Bra Joe and Bra Dan is that they do not know in which story they are appearing until they have already said their lines and performed their deeds. They disarm a man on a Friday night, take him to the police station, write a witness statement, hand the statement over to the detective on duty, and leave. Walking home from the station, they do not yet know to whom they have given that statement: to the police of the first story, or to the market player in the second story. That is something they will discover in the weeks to come.

As they lead their phalanxes of elderly men through the streets of Alexandra, they walk with one foot in either story. They are, alternatively, citizens bringing policing to their community, and members of a self-defence group that dare not stop its work of self-defence.

*

The second story is familiar to millions. It is the story of black, urban South Africa under white rule.

In its more caricatured versions, the history of South Africa told by its liberation movements is one of a black nation united by its common oppression. Despite the fact that generations of South Africans were forced to live wretched and uprooted lives, the masses have always risen above their circumstances to claim a dignity that inheres in all oppressed people, the story goes. True, they never had a police force they could call their own; there was never a state agency active in black urban life that kept the peace. But the people were united *against* a hostile police force, and with that unity came organic forms of peacekeeping; popular structures that kept order and resolved disputes were plucked from pre-colonial history, from sheer decency, from the exigencies of staying united in the face of attack.

In this scheme of things, black South Africa at the beginning of the democratic era was a nation waiting for a legitimate police force. It was, and always had been, ready, when circumstances permitted, to lay down arms and give consent to being policed by an agency that would enforce the community's own norms. According to this story, the darker aspects of the tale I have told thus far – the crowds that attack police when police are outnumbered; the buying and selling of police services on open markets – these are residues from another time, signs that there is always an interregnum between the old and the new.

This tale about the past is too dreamy and too fanciful. Difficult times do not produce gentility, as if there is an angel hovering over the world of the oppressed. Under white rule, black urban South Africa was a bewilderingly diverse and cosmopolitan place. Wave upon wave of new people arrived; they were far from home and from safety, and the place to which they had come was governed by hostile people. It was a world of strangers and of estrangement. If ever there was a place that cried out for an agency that enforced the common rules of a modus vivendi, an agency before which everyone else abrogated their entitlement to use force, it was the cities of white-ruled South Africa.

Yet such an agency did not exist, not in black spaces at any rate. And so among the associations South Africa's cacophony of urban residents formed were associations of defence, protection and violence. And since the members of these associations were not given uniforms and salaries by the state, and since theirs was not to enforce a set of rules to which everyone had agreed to submit, they also used their capacity for violence to sell protection, to control space, to regulate access to resources such as housing, and to defend markets for such commodities as alcohol, sex and transport. Rival associations would inevitably clash over markets and over turf. In this world, the white state's police force was sometimes a menace to be thwarted, sometimes a useful tool to be bought, but never a neutral enforcer of common rules.

*

Dan Sibanda and Joe Mukwali's story echoes throughout the histories, memoirs and fiction that document black urban life under white rule. In some cases, the echoes are so true that only the names of people and organisations differ.

In Soweto in the early 1970s, the groups of elderly residents who patrolled the streets in search of men to disarm were called *makgotla*. Their adversaries were often their own sons: gangs of Soweto youths who controlled the space around the township's bus and train thoroughfares, and made a living robbing commuters moving to and from work.

In the late 1940s and early 1950s, in Western Native Township and Sophiatown on the western fringes of Johannesburg, the elderly men pacing the streets were called the Civilian Guard, a formal body constituted through the local ratepayers association. Just like the current Sector Four Patrol Group in Alexandra, their aim, writes the historian Gary Kynoch in his book, *We Are Fighting the World*, was to 'patrol the streets, to disarm people found with weapons, and to turn offenders over to the police.' When it was formed in 1951 it was about 1 000-strong. As the patrol consolidated, it began moving its activities south, to the adjacent township of Newclare, which is when its troubles began.

Newclare was dominated at the time by Basotho migrants, who had their own self-protective organisation, the infamous Marashea, or Russians. It is no exaggeration to say that the Russians controlled Newclare at this time. They provided security, regulated commercial life, and controlled access to housing.

'Living under Russian rule would have appealed to many migrant Basotho,' Kynoch writes. Many were in Johannesburg illegally, and thus could not work or move freely through the city. The Russian-controlled zone of Newclare was thus something of a refuge, in which 'these migrants were able to engage in income-generating activities that allowed them to scratch out a living.' Moreover, the Russians allocated housing and other infrastructure, and the protected Basotho from violence: 'The rudimentary Russian code,' Kynoch writes, 'dictated that ordinary residents should not be robbed, old people were to be respected, and only tsotsis and members of rival ethnic groups ... were legitimate targets for assault.'

When the Civilian Guard moved south into Newclare disarming residents, the Marashea obviously understood this as a grave threat. 'The Russians retaliated with a vengeance and attacked patrolling Guards on Christmas Day, 1951,' Kynoch writes. 'In the ensuing battle eight men were killed and 20 injured.'

At this stage, the police entered the story. The

Russians were an old and experienced organisation, and while they were extremely wary of the police force, they had long ago learned to manipulate it when they could. The Civilian Guard was allied to the ANC. The Russians began to use this fact against them, spinning a story to the government that the Guards were a menacing communist presence in Newclare, the Russians themselves a bastion of the established order. The police joined the conflict on the Russians' side, destroying the Civilian Guards, only to turn on the Russians themselves once their common foe was gone.

*

There is a common thread linking the Guards of Newclare in 1951 to the Sector Four Patrol of Alexandra in 2007. Both try to become adjuncts to the police service of their day. But neither operates in a context that has a police service in the most meaningful sense of the term.

Like Newclare, indeed, even more so, contemporary Alex is a bewilderingly heterogeneous place. Poor people are moving to Johannesburg in vast numbers from all directions, and Alex, with its proximity to central Johannesburg, and its long-established networks of migrants, is prime space. There are the Zulu-speakers who live in Madala and Nobuhle hostels, and who organise their own protection; the police dare not enter these places except in large numbers in the early hours of the morning. On the banks of the Jukskei River there is a dense shack settlement run by groups of Mozambican immigrants, and along Eighth Avenue, in the gaps between the clusters of apartment blocks, there are shack settlements predominated by recent Zimbabwean immigrants. Each of these spheres of Alex has an associational life, and all these associations perform functions that protect kith and kin.

Even in the yards of old Alex, where Bra Dan has spent half a century, life is not simple. There are people there who earn money buying and selling parts of hijacked vehicles and goods taken in house robberies; such people do not appreciate Bra Dan's patrols. There is also, in and around the yards, a generation of unemployed youths, many of whom are making a living by being armed, by moving goods in and out of Alex, by being unpoliced.

So wherever the Sector Four Patrol Group goes, it upsets arrangements, networks and institutions at the heart of people's lives, and people fight back. Among the weapons they wield is their capacity to corrupt members of the local police station.

Bra Dan and Bra Joe are at once part of a new story as well as players in the same old story. In the realm of security, the transition from apartheid has both gone a long way, and has barely begun. The current police service has not yet garnered sufficient moral authority to rise above the logic of the past in which violence was traded in return for money, for sympathy and for tactical co-operation. In the following two chapters, I explore one aspect of why not. I examine what the police looked like in the eyes of township residents when they returned to the townships in the early 1990s.

Mtutu

To ask what township residents saw when they watched cops return to daily life in the early 1990s is to locate the police in collective memory. Each and every police officer who walked or drove through the townships during this country's transition to democracy bore the marks of history: the history of what the police meant in general, and of what he himself was doing a few years earlier.

And so, we must reel back in time. The best place to start, I think, is the early 1970s, on the eve of the great uprisings of 1976, after which nothing would ever be the same, particularly not the police.

*

In late June 2007, I meet with the veteran Soweto writer Mtutuzeli Matshoba to discuss the past. Matshoba became famous in the late 1970s with the publication of his book of short stories, *Call Me Not a Man and Other Stories*, one of the most eloquent and thoughtful documents of 1970s Soweto. I have phoned him a week earlier and asked whether he'd give some thought to his experiences of the police during those times. We sit down together at a coffee shop table on Mary Fitzgerald Square in downtown Johannesburg on a Thursday mid-morning, and it is well into the afternoon by the time Mtutu has done talking.

Mtutu was born in 1950 and grew up in Mzimhlophe, Soweto. In the story he begins to tell me, he is in his late teens and early twenties, and we are thus in the immediate, pre-76 era. The police his tongue crafts is not a monolith, but a complicated filigree of people and organisations, some dark, others benign, still others a treacherously mercurial combination of both.

The first category of police Mtutu recalls are the 'blackjacks', so called because they clung to you and didn't let go. They were not members of the nationally run South African Police, did not carry guns, and had received only a cursory training. They were municipal police officers whose main task was to enforce the pass laws, and they were the most visible police presence on the township streets.

'They were the raiders,' Mtutu recalls, 'the ones who would pound on the door at 2 am to see if the house was harbouring anyone who did not have a stamp in his passbook permitting him to be in the urban areas. We called them '*Gqoka Sihamb*' – "put on clothes, let's go." And that was their power; one moment you were asleep in your bed, the next you were packing your toothbrush.

'It is almost magical how powerful those pass laws were,' Mtutu continues. 'A blackjack walks in to see your pass, he takes it, puts it in his pocket, and you must follow him. He gathers a few people like this. One cop followed by 12 people. He is on his bicycle, the 12 are jogging to keep up with him. He stops to see this and that one, to drink and to visit, and the 12 wait for him outside. Then he gets on his bicycle again, and the 12 jog again to keep up with him.

'He has your passbook in his pocket, and you are nothing without it. You cannot *be* without your passbook. You must follow him.'

As Mtutu tells me this story, my mind wanders to an evening in February 2004, when I accompanied a Saturday-night patrol in the West Rand township of Kagiso. At about 9 pm, a flurry of domestic violence complaints came over the radio: about a dozen in as many minutes. The two cops I was accompanying were incensed. They went from one complaint to the next, walked into each house, grabbed a man by the scruff of his collar, threw him into the back of their van, and booked him into the station cells on the grounds of public drunkenness. The cops that night were like a veterinary service bringing in a haul of rabid dogs to be put down.

'There was no time to ask questions,' one of them explained to me. 'And if we leave a man in the house and he kills his wife two hours later, we are in big trouble.'

Most of the men thrown into the back of the van that night were drunk, but none was in public: each was in his own home. And yet not one of them questioned the legality of his arrest. After nearly a decade of democracy, each assumed that the cops had every right drag him out of his home and throw him in prison. Mtutu's blackjacks, it seems to me, cast a very long shadow into the future; some forms of submission are so habitual they take generations to die.

*

The category of police about whom Mtutu speaks most vividly are the reservists.

'When influx control began to crack,' Mtutu recalls, 'they signed up people they called reservists. Their job was to look for people without passes. Anyone who was prepared to do it could do it. No knowledge of policing was required.

'There was a man in Mzimhlophe called Skosana. He was a reservist with criminal tendencies. He preyed on hostel dwellers; he forced them to pay him weekly bribes so that he would not arrest them. He and his friends were like hell lions sauntering down the street, sweeping anything that was male. You couldn't talk back. If you did, they would plant a knife on you that they had taken from someone who had bribed them earlier. It was legalised extortion.'

As a young man, Mtutu lived in his parents' house opposite Mzimhlophe railway station. His yard looked directly onto the station platform.

'From my mom's place,' he recalls, 'I could stand and watch the reign of terror Skosana and his friends unleashed on rail commuters. They prowled around the street outside the station like hungry animals. There were people standing there in terror. Anyone without a pass was marooned on the station until they left.'

The Matshobas got to know the reservists well.

'My father never allowed them to step foot in his yard,' Mtutu continues. 'He would pick up a stick and challenge them. He despised them. I learned that from him. I did the same as him. People fled into our yard from the station and I would keep the reservists out.

'So they knew me: they knew me as somebody whom

they must teach a lesson. If they meet me in the street I am in for trouble. If I run into them, I must fight. The police-proper could kill you. The reservists only had sticks or truncheons. I must fight them with my hands.

'One afternoon at the station, they stopped to search me. They found nothing. They slapped me across the face. I fought them. And it so happened that while we were tussling, my mother and her friends appeared on the platform. They were coming home from work. My mother was a nurse.

'You will not believe what they did. They joined in on the side of the reservists and laid into me with their handbags. "How can you fight the law like that?" my mother shouted. "These men are working for the law. What trouble are you trying to bring?"'

Of all the stories Mtutu tells me, this one sticks the most. Its meaning is slippery, the lessons it teaches complicated and unclear. These men were common thugs to whom the state had given uniforms and sticks. Those who watched them going about their business were witness to the fact that under white rule, black life was at bottom lawless, that the most depraved members of a community were given licence to terrorise the meek.

And yet when she saw them beating her son, Mtutu's mother joined in with her handbag. Why? On the one hand, she was acknowledging that these thugs were dangerous, that the state itself was lethally dangerous, that in this world her son ought to keep his head down. And so her anger was maybe the displaced expression of a great fear.

But perhaps something else was going on. Mrs Matshoba's son was a young man at a time when many young men were running wild. This was the era of the Hazels Gang, which enforced a reign of terror on the trains, fleecing entire carriages of their pay packets and their watches during the seemingly endless stretch between stations. It was a time of notorious thuggery, and without a decent police service, there was nothing to keep it in check. So perhaps Mrs Matshoba was expressing a hunger for authority, for men in uniform who would throw a cordon around her son's life to stop him from straying. That she had to invest this wish in the thugs who menaced the Mzimhlophe station platform has about it a feeling of pathos; in this deformed world, order and authority became confused with state violence and depravity.

*

This hunger for order is prevalent in a great deal of testimony from the 1970s and 1980s. Thabo Mopasi, who has lived in Alex all his life, and whom I interviewed in August 2007, recalls that throughout his early childhood in the mid-1970s, his mother resided in Alex illegally and had to play a constant game of cat and mouse. It was a bizarre situation. Her husband had a pass, and her children's residence in Alexandra was legal, but hers wasn't because she had no work.

'The municipal police,' Thabo recalls, 'the ones with the khaki uniforms, would pound on our door at four o'clock in the morning. My mother, my younger brother and I would run out the back. We lived on 13th Avenue. We would run across to another yard, climb a fence into 12th Avenue, and hide in my father's car, which was always parked in a garage on that street. From the garage, we heard the police kicking doors. If they caught my mom she would have 72 hours to leave Johannesburg.'

Thabo and I are sitting on a bench in the grounds of his old high school in Alex. Seamlessly, as if still telling the same story about the same subject, he drops the municipal police and begins speaking of his father.

'It is frightening to think how much cruelty there was in my life,' he says. 'I did not know better then. I thought it was normal. My dad would beat my mother very heavily when he was drinking: every weekend, one after the other. Women did not go to the police to report battery in those days. It was unheard of. It was better to run away to neighbours, or to my uncle on 15th Avenue, and even that was not a thing she would do lightly.'

'At some point,' he continues, 'we moved from our old place to a communal yard. We rented a room from the family that owned the main house in the yard. There was a disco in the back owned by that family. It had rocking music that went on through the night; every Friday, every Saturday, their house was full. The police station was far away in Wynberg. It was never thought of that there were bylaws that must be enforced around a shebeen: drinking in the streets, closing hours, the carrying of dangerous weapons. Weekend after weekend, we would wake up in the morning and somebody has been stabbed, somebody has been killed, whether walking home from a shebeen, or outside a shebeen, or even at the very shebeen in our own yard.'

Thabo's is a story told in hindsight. He could not have expressed it this way when he was a child, but looking back now, he sees a boy searching for a thread with which to stitch up the holes through which violence is dripping into his life. His father's violence against his mother; the drunken violence that threatened every weekend in his yard: in his retrospective story, he is a boy yearning for a police force, an agency that is called when things get out of control. As it was, the police took their place as yet another source of instability in his life; they were indeed the supreme menace, for if she was not always on her toes, they would take his mother from him.

*

The picture is not quite as stark as that, not quite as simple. There was, in the pre-76 era, a particularly interesting township figure who was both a policeman and a genuine source of authority. He was often a neighbour, sometimes a relative. He wore a jacket and tie. More often than not, he left the township early in the morning and only returned in the evening, for he worked elsewhere. He was a Criminal Investigation Division man, a detective, and he really was a Janus-faced being: he was mistrusted because he was employed by the police, feared because he worked with violence, but deeply respected because white-collar professionals who worked with pen and paper were considered middle class and were looked up to.

Luyanda Msumza was in his mid-twenties in the pre-76 era, and a resident of Ginsberg in the Eastern Cape. His father was a CID man. Some months after the murder of Steve Biko in 1977, Luyanda was to join the Pan Africanist Congress in exile, and he would not see his father again until his return to South Africa in 1990. But in the pre-76 era, he remembers his father as a figure of great authority in Ginsberg.

'People would literally queue outside our door waiting to ask him to settle problems,' Luyanda recalls when I interview him in July 2007. 'Whether it was because of family disputes, or a man beating his wife, or a fight between neighbours: people came to him when there was trouble.

'You see, political consciousness was the preserve of a very, very few people in those days. Never mind that my father was a policeman in the apartheid state; people saw him as working in a better job than others, and that made him a moral leader of sorts. If he solves a theft, it is a theft in the black community he has solved. He has brought something to the community.

'I remember once, my father and I were standing outside the house, and a youngster walked past carrying a big pot on his head. "Where are you taking that?" my father asked him. "Put it down and leave it here." The boy never came back. It turned out that the pot belonged to a friend of my mother's. When she saw it, she said: "Oh, I thought I lost that a few months ago."'

I have been to Luyanda's ancestral home in Lusikisiki, where a portrait of his father stares down into the living room from its place on the wall high up near the ceiling. He has a grey, handlebar moustache and in his eyes an air of authority so severe I fear for the one behind the camera's lens. He strikes me as a blunt and intimidating figure. I tell Luyanda so.

'He was not a non-violent man,' Luyanda replies dryly. 'But then again, the township has never been a non-violent place. People stab and kill in the township all the time. You bury, you go on. People have always lived with this; sometimes there is even intermarriage between the families of the murderer and the deceased.

'In any event, my father only killed one person, a notorious man called Khotso whom everyone feared, and whom my father shot when he went to arrest Khotso and Khotso attacked him. When I say my dad was not non-violent, I mean that he would discipline boys who engaged in petty theft with a very heavy stick. There are still old men around who remember vividly the beatings they got from my father in the 1950s and 1960s. There was some thoughtfulness behind his violence, some care about the community he lived in. He abhorred people who sent children to reformatories. Why waste a young life, he would ask. Rather give them a good hiding and let them go on being children.'

*

Every township resident of the early 1970s has stories about the CID man in his neighbourhood. Mtutuzeli Matshoba's was a large-framed, middle-aged man named Mqaba, who worked for the notorious Brixton Murder and Robbery Squad.

Mtutu delights in paradox. The Mqaba he conjures is a dizzying, dangerous combination: feared and yet constantly solicited, mistrusted and yet called upon to intercede in difficult, private matters.

'We were scared of him because he was CID,' Mtutu recalls, 'and especially because he was Brixton. But as far as I know, he never laid a finger on anyone in Mzimhlophe. It was a question of his aura. We used to say that there in Brixton at the Murder and Robbery Squad, they have what is called the *waarheid kamer*, the "truth room". They take you in there, and you do not leave until you have told the truth.'

'And yet, despite that,' I suggest, 'he was also a morally elevated figure?'

'Well, he was a traditional person, a Xhosa man with a big family. People would go to him to resolve disputes. It was an informal role. If somebody was bullying you, you would go to Mqaba and he would intercede. Unless the crime became more serious; if a gun was involved, then a proper police investigation would proceed.'

'But make no mistake,' Mtutu continues, 'Mqaba may have been "morally elevated", as you say, but at the same time he was morally tarnished. Somebody like him could never be wholly trusted.

'Elias Motsoaledi's family lived 200 metres down the street. I played with his sons before he went to Robben Island. After he went, playing with them would be questioned by the family. My mother did not want me to go there. They were isolated. His wife had no friends.

'What was it, exactly, that people were afraid of? If I played with the Motsoaledi kids, who would tell the regime? After all, there were no whites in Mzimhlophe. It was the Mqabas of the world who would whisper in the regime's ear. Even if the regime came to get you, no white cop could tell where you stay. It took black *impimpis* on the inside to show them where to go.'

*

Among the boys in Mzimhlophe when Mtutu was growing up was a notorious member of the Hazels Gang. He was coy when I interviewed him in July 2007, and we agreed that I would refer to him here as Bra X. He told me that he and other Hazels would ambush the vans that came out of Mzimhlophe, having delivered their loads of meat and cigarettes. They would hold up the drivers at knifepoint, and make off with the cash.

'In broad daylight,' he tells me, raising his eyebrows, as if the very idea of it still astonishes him now. 'Our fathers feared us. Grown men, middle-aged men feared us. The only one who had no fear for us at all was Mqaba.'

'That man was a father/policeman,' he says, 'a father you hate and love. We would be standing around on the corner planning trouble, and he would come and confront us. "Boys," he'd say. "I know why you are here, and I also know your parents." Then he would take out his wallet. "I am going to give you one rand each, to keep you out of trouble. And now you must move off. If I hear that you have made trouble, I will personally raid your houses."'

A gruff Father Christmas, and yet also a snitch who may betray your family to the apartheid regime. Such was the slippery, dangerous character of authority in urban black life under white rule.

Driving through townships with cops more than three decades later, it is clear that this treacherous ambivalence persists: these uniformed figures called into the intimacy of homes to order the chaos of private lives, and yet who avoid crowds for fear of being turned upon.

*

Mqaba was not the only sort of township CID man in the early 1970s. Others were out-and-out thugs who used their positions in the South African Police to trade violence and favours in the township underworld. The thug-detective of Mzimhlophe during Mtutu's era was the notorious Hlubi, whose killer would one day be given a medal for his deed by Nelson Mandela before an audience of thousands.

'Hlubi was one of three notorious cops in Mzimhlophe,' Mtutu tells me. 'The second I grew up with: Rampae. The third was Senoamadi, which means bloodsucker; I don't know whether that was his real name, or a nickname.

'The three of them were the terror of Soweto at one stage. They were all Brixton. All pushed weights. All were huge and strong. Their names sent shivers down spines. Especially Hlubi: he would find someone smoking dagga in the street and shoot him, then say he was resisting arrest. He would walk into a shebeen, grab someone, and drag him out into the street by his foot.

'This is very much a pre-76 story I am telling you: until 16 June 1976, people would not dare to resist a policeman.'

Mtutu himself brushed against Hlubi once, and the encounter might have cost him his life.

'There was a shebeener in our street called Lilian. She was in love with Dr Shivia; they were lovers for some time. Then she had an affair with Hlubi. He kicked Shivia out at gunpoint. He was that sort of man; he was very, very dangerous, and there was no law above him, nothing to stop him.

'I walked past the shebeen one day to find that a

friend of mine was being beaten up. I grabbed a stick and defended him. These people beating up my friend were connections of Hlubi's. Word got to him; we are interfering with his girlfriend's shebeen.

'I went immediately back to Fort Hare, back to university. I had been in Soweto only for the university holidays. Now Hlubi was out looking for me, and got very angry that he could not find me. He and Rampai grabbed a friend of mine off the street and took him to Brixton. They wanted to know where I was. My friend would not tell them. So they handcuffed him, and they put one of those contracting rubber bands they use in tyres around his head. It is wrapped tight around his head, and then it contracts. It made him deaf in one ear. He is deaf in that ear to this day.'

*

On 16 June 1976, a new world was born. There would no longer be much place for Mqaba and his ilk, the frightening but respectable neighbourhood CID man to whom people took their problems. The post-76 era was made for the Hlubis of this world; he would kill and be killed in turn.

Police as allergy

Bra X, the Hazels gangster who grew up in Mtutuzeli Matshoba's neighbourhood, claims to have been an eyewitness to a murder on 16 June 1976. Three whites were pulled from their cars and killed by the crowds on that day; Bra X says he stood over the corpse of one of them.

'He just happened to come by at a very unfortunate moment,' Bra X says. 'I was in a group marching towards Uncle Tom's Hall. We had just seen the police, and the crowd was angry and excited. And this old white man came driving his municipal van. He was dragged from his car, stoned to death, and put in a dustbin.

'His car keys were left lying on the ground. I picked them up. The key ring was attached to a yellow tag. I will never forget that. Until that day, I had taken whites as superior beings, untouchable. But he was crushed, like he was nothing.

'Everyone was feeling the same as me. The section of the crowd that had watched him die went mad. We started looking for vehicles to hijack. We wanted to use them as battering rams to break into a bottle store.'

Bra X was a gangster. He was no friend of the students who led the uprising. His relation to the anti-apartheid movement in the coming years would sway between deep suspicion and overt hostility. But his sense that something momentous had happened, that nothing would ever mean what it used to, from the bodily integrity of a white man to the shop front of a bottle store, is a good index of what happened that day. Among the things that had changed forever was Soweto's relationship with its police.

'Before 1976,' Mtutuzeli tells me, 'we felt that the cops were untouchable. To kill a cop was beyond the imagination. When 76 happened, the riot police came out, and that was the last straw. The cops were there flushing people out, shooting people, terrorising people, and all that fear we had for them was replaced by fury. It is almost magical how it happened. One day, they are the most feared creatures on this planet, the next, it is not safe for them to walk through the township.

'There was no ordinary policing after June 1976. The township policemen could no longer live at home. They moved into barracks. As for the blackjacks, policing the pass laws was now unthinkable. The blackjacks disappeared, and when they returned a little while later, they were now the Green Beans, because they wore green pants now, and they had a very different function. There was no more *gqoka sihambe*: they could no longer come to your house in the middle of the night and take your pass; you would not find people jogging after their bicycles. They retreated. When they came back, it was as security guards. They guarded the buildings we were attacking, the council offices, the bottle stores. And they could only do so in large numbers. If you saw three or four them on their own, you would attack them.

'I remember in September 1976, two friends and I were sitting in a train. A cop, wearing his uniform, walked into the carriage. We were drunk. We looked at one another, and each knew exactly what the other was thinking. Without exchanging a word, we attacked him and stripped him of his uniform. We shared it among ourselves. I think I was wearing the cap. It was amazing. I had never attacked anyone in a carriage before, let alone a policeman. Three months earlier, it would have been utterly unthinkable.

'So we are on the streets with various parts of his uniform, walking brazenly like we owned the world. We walked into a shebeen, with me wearing this police cap, others wearing other parts of the uniform. The shebeen queen saw us, and her face clouded over. She took us aside. She made us burn that uniform. She was scared for us. She wanted to protect us. But I must tell you, it was very satisfying seeing that policeman's fear.'

One by one, the respective fates of the cops who had peopled Mtutu's pre-76 world darkened. Mqaba was in late middle age now, and two of his sons had followed him into the CID.

'One of them,' Mtutu recalls, 'was famous in Soweto because he was a very good footballer. He came close to turning professional. On the day that Zindzi Mandela spoke at Orlando Stadium, a very tense day, he was seen firing teargas. And that was the end for him. He came to the township some time after that to visit his mother, and he was beaten to a pulp, his gun taken from him.'

As for Skosana, the reservist who had terrorised the commuters on the Mzimhlophe station platform, and had forced Mzimhlophe hostel residents to pay him protection money: 'He believed that nothing had changed,' Mtutu says. 'He believed that he could go on as he had always done. He used to walk into the hostel to collect payment. One day, he went there as usual, and they told him to go to a certain room to get his money. When he walked through the door, the Zulu guys in there ambushed him and hacked him to death.

'There was celebration in Mzimhlophe that day, and it is interesting that there was, because the hostel dwellers were no friends of the uprising. They were very much against the uprising.'

Hlubi's post-1976 fate was another story entirely. In Mtutu's memory, Hlubi is the embodiment of the evil unleashed by the state's response to the uprisings.

'In retrospect,' Mtutu says, 'it seems that the post-76 era was designed especially for Hlubi and his ilk. It was then that he came to prominence and his infamy grew. A green Chevrolet surfaced a few days after 16 June 1976. Hlubi and other marksmen were inside. It was said that they betted a bottle of whiskey on who would take someone out that day. A boy called Makhosini, a schoolboy in the neighbourhood, was playing with some other kids on a patch of grass. The green Chev came past, shots rang out, they hit the ground. One of them did not get up again. There was a clean little hole in the back of his head. They turned him over, and his face was a mess. Makhosini couldn't go to school again. He was too traumatised. He was taken to school in Natal, and we did not see him for a long time.'

When the Truth and Reconciliation Commission heard evidence from Soweto victims of state repression two decades later, Hlubi's name was all over the place. One after the other, former student activists said that they came back to Soweto after spending time in detention, and then immediately heard that Hlubi was looking for them; they went into hiding or fled the country. The most thuggish, criminally minded Mzimhlophe cop of the pre-76 era became the frontman of lethal political policing.

Everybody, Mtutu tells me, remembers where they were and what they were doing when they heard that Hlubi was dead.

'It was 1978,' Mtutu recalls. 'My brother had been detained for his activities some time earlier and taken to Port Elizabeth. We went to a lawyer, the best for these circumstances. His name was David Botha. He never lost a case. We fought a hard campaign. Helen Suzman got involved. We eventually secured my brother's release.

'The arrangement was that he was to be transferred from Port Elizabeth to Protea Police Station and released from there. The family went to the station to get him. When we arrived, I had a copy of that day's *Sowetan* rolled up under my arm. There was a riot cop outside the station standing there in full riot gear. He asked to see my newspaper. I opened it up. The headline on the front page read: "Hlubi Shot". The riot cop looked at it and said: "Actually, he is dead. He was rushed to Baragwanath when he was shot, and the nurses killed him."

'He had been ambushed outside his house the previous day, and badly wounded. The rumour going around the township, the one the riot cop had obviously heard, was that when he arrived at the hospital all shot up, the nurses looked at one another and said: "We will have no men left if he survives."

'By the next day, every criminal in Soweto was claiming responsibility for his death. There were so many myths, so many stories, one did not know where to turn. And then, many years later, in the early 1990s, there is a very big rally at Orlando Stadium to close down Umkhonto we Sizwe. It is in the middle of the negotiation process, and MK is closing down. Mandela is making a speech, and he calls Jabu Masina onto the stage, a boy who had joined MK after 76, and he gives him a medal for having killed the dreaded Hlubi.'

*

These excerpts from Mtutu's memory are painterly and impressionistic; he jumps from 1976 to Zindzi Mandela's famous stadium speech nearly a decade later in which she delivered a message from her jailed father, back to Hlubi's death in the late 1970s. He is drawing from feeling and from memory in order to describe what it was he saw when ordinary policing returned to Soweto in the early 1990s.

There were of course vast differences in ordinary people's experiences of the police across both time and space. In much of the country, it was not in 1976, but only during the latter stages of the 1984-86 uprisings that the police were forced to suspend normal activity. In Alex, for instance, cops like Captain R lived in their homes in the township until February 1986.

But by the end of that year, the uprisings had touched

just about every black urban space across South Africa, as well as dozens of rural towns; millions of black South Africans had seen the suspension of policing in their neighbourhoods, and its replacement by quasi-military riot control. When the police are pushed out of a residential space in a more or less organised rebellion, the question of order immediately arises. The rebellion needs to police itself, its enemies, and the space to which it wants to lay claim. As much attention must be paid to internal definition as to external attack.

This is universally true of uprisings, but since we have been talking of Soweto 1976, we may as well use it as an example. That gangsters like Bra X joined the crowds on 16 June is something of an irony. The uprisings were organised by a movement of high school students. Gangsters and school-goers were sworn enemies in Soweto throughout the early 1970s. Tsotsis would attack teenaged boys and girls on their way to and from school, and schools would organise themselves into permanent structures of self-defence.

And so when gangsters joined the uprisings *en masse* on June 16, student leaders wondered whether it was a blessing or a curse: an opportunity to bring thousands of uneducated youths into politics, or a corruption of politics by thugs.

The choice was made for them. In August 1976, Zulu migrants resident at Mzimhlophe Hostel turned violently on Soweto youths, for they saw little distinction between the students and gangsters, and gangsters were their historical enemies. The students also risked losing the support of their parents and of workers, for whom the involvement of tsotsis in the uprisings was repugnant. And so the student leaders chose migrants and elders over gangsters, and began to perform policing functions themselves. They came to 'recognise the importance of crime as a popular grievance', the historian Clive Glaser observes in his book *Bo-Tsotsi*. 'By actively combating crime, the Soweto Students Representative Council could demonstrate its authority and its community responsibility.'

Every uprising must play this policing function, defining friend and enemy, establishing a determinate order. Whether it does so benignly or with viciousness is not the point here. The issue is that its legitimacy as enforcer can never be universally accepted. In cosmopolitan and underpoliced urban spaces like those of late twentieth-century South Africa, the emergence of a new enforcer is considered by some to be a grave threat. The Soweto students were able to choose to some extent who was to be friend and who enemy, but they couldn't choose to have no enemies in their neighbourhoods at all.

To see what I mean, recall the conflict in the early 1950s between the Civilian Guards of Sophiatown and the Basotho migrant gang, the Russians, of Newclare. In the absence of normal policing, the Russians had long come to play a host of regulatory functions in the Basotho-dominant neighbourhood of Newclare, providing safety in return for monthly payments, gatekeeping access to housing, securing a sheltered space in which Basotho who were in Johannesburg illegally could live and work.

When the Civilian Guards – a body of decent people intent on bringing policing to their neighbourhood –

moved their patrols south from Sophiatown into Newclare, the entire system the Russians controlled was threatened. And so they fought back.

With the 1980s uprisings, what happened in Newclare on a local scale more than three decades earlier, began happening in black residential space all over the country. Anyone who, like the Russians, was in the business of providing security, felt deeply threatened by the youths on the streets, the people's courts, the street committees. So many age-old tensions became open conflicts: between old and young, hostel residents and township residents, newly urbanised shack dwellers and people in formal houses.

The apartheid government fought the uprising by exploiting these tensions. From renegade youth gangs in Soweto, to recently urbanised Xhosa traditionalists in Crossroads outside Cape Town, to generations-old Zulu hostel organisations across the Reef, to the remnants of the Russians gang in mining settlements – everywhere, the government found people threatened by the uprisings and only too happy to enter into an alliance with the white state, just as the Russians of Newclare had done in 1951.

The 1980s were wild days. And there was a paradox about their wildness. On the one hand, the most prevalent discourse of the times spoke of a popular black uprising against apartheid. Never had black South Africa appeared so unified. Never had so many been heard to speak with such authority on behalf of everyone. And yet in the fabric of township life, much of the authority that had once existed – the authority that came with wearing a suit and tie, or with being old, or with commanding a bounded space like a hostel – had unthreaded. And so every assertion of authority was in fact a provocation.

I discuss with Bra X his memories of the street committees and the people's courts that arose in Mzimhlophe in the mid-1980s. As a gangster, he was a natural enemy of the people who staffed these structures, and he is at pains to show me that their self-ordained role of bringing order and justice to the township was unrealisable.

'There were a couple of older, wiser people in the street committees,' he concedes, 'but power corrupts very quickly. X is interested in Y's wife. Y is on the street committee. Soon, the committee comes to X's house, drags him into the street and flogs him, and what is happening is that Y is using the whole committee to police his wife. And somebody in the street committee would inevitably have a relative without a house. Somebody would be driven out of their house to make way for the relative. You can say that I was a criminal, and the street committee anti-criminal, but life is very dynamic. You can be a Christian, me an atheist, but if you watch us carefully we are doing the same thing.'

Bra X warms to his topic. 'In fact, I can say,' he continues, 'that the presence of the street committees taught me that I could use violence in ways I had never thought of before. Take an example. My mom's house in White City had a problem. This was in 1986. She had given her house over to her nephew and his wife. The councillor for the area, Councillor Maseko, saw that there were now other people in the house. She started serving notices: the house must be vacated.

'My mom never discussed this with me. My son is a criminal, she thought; he has no interest in domestic problems. I found out through others. I confronted my mom. She explained what had happened. I said, okay, give me this councillor's house number.

'I went there and knocked on the door. They let me in. Councillor Maseko was not there. Fortunately, her mom was there. I took out my gun and I said to her: "If you still love your child, tell her not to go to my mother's house anymore. Actually, she should have come home to find your corpse. But you are not part of this. I will leave you to tell the story." As soon as my foot was out her door, she started screaming. But to this day, that house has had no further problems.

'As I was walking away, I thought to myself: "I have learned something from the street committees. Before, I used to think that the only use for my gun was to take money and goods from people. From the street committees I have learned that you can use your gun to secure your family in their accommodation."'

In the 1980s, the police were obviously in no position to tame this growing wildness. But they were in a position to inflame it, which is precisely what they did. In Mtutu's and Bra X's world, a criminal gang called Kabasa turned against the civic and youth organisations in Orlando East, and the police helped them to pick out targets. Across the country, alliances between police and a diverse assortment of vigilantes mushroomed. Fighting between groups laying rival claims to order and to self-protection had never been more vicious, and the police had never been more active in encouraging violent conflict.

The nature and extent of these poisonous alliances between police and township residents remains a matter of fierce dispute to this day. How much of the police organisation was involved, how much wasn't, how much national leaders and local commanders knew: these questions are destined never to be settled.

The point is that any semblance of ordinary policing had vanished, and the meaning of police was thus simplified and reduced. As a body, the police force was disgraced; it was associated above all with social malignancy, and its members inspired disgust. In the early 1970s, the meaning of police was relatively complex. There were thugs like Hlubi and ersatz police-criminals like Skosana, but there was also the aura of the white-collared CID, the respect earned by men like Mqaba and Luyanda Msumza's detective father. That world was gone by the late 80s. The police was now considered a force of sheer destructiveness, a kind of social allergy that inflamed existing tensions and agitated animosities.

It is seldom noted that the meaning of police had a long history under white rule, and that its reputation was at its most vile only moments before it was asked to step forward and police a new democratic society.

I ask Mtutu what went through his mind when he watched police officers move through Soweto after the ANC came home. By way of reply, he begins telling a story located a little earlier, in the late 1980s.

'Some time in the early or mid-1980s,' he tells me, 'special police housing was built in Protea North for the ones who could no longer live in their own homes. I went there four or five years after it was established.

The area was full of them: every fourth house, a cop lived in.

'I arrived with my ANC bravado. I was wearing a very provocative T-shirt, if I recall. I found that the police were mixing with ordinary people. Cops in uniform drinking with civilians. I was on tenterhooks. This was not something I had seen for a very long time.

'And so I sat and drank, and I watched very carefully. It did not take long to see that these cops were trying to ingratiate themselves to people. They had lost any semblance of dignity. It was if they were holding begging bowls. It was quite a pathetic sight.

'And really, although that was 20 years ago, and 20 years is a long time, between then and now is really a continuum. Yes, the young ones now are not the same people as the old ones who were forced to run away, but they have inherited the same burden.

'What has changed is that they can freely socialise, they can have girlfriends in the community. But they are the rubbish of the township. They are the lowlife.

'When the street committees were removed, the new ANC government tried to replace them with Community Policing Forums, but they were useless because the cops were not respected. They are subservient in Soweto. They get sworn at. People talk back at them. People spit at them. Female cops are more respected. I don't know why. I would rather report something to a female cop. Most male cops are alcoholics. They smoke dagga. Lots are involved in robbery. They aspire to be clever, street-smart, enforcers for big criminals. They can squash a case. That is what they are good for. It is a racket. Their function in life is to collect money. A guy is drinking beer in the street, or pissing in the street, they take money. A guy has a beaten-up car, they take money. If you are a black person and you encounter a cop, you must be ready with graft. It doesn't matter the circumstances. To take a case seriously, or to make a case go away – anything.

'Take my tax problems. I'm only learning now it's a serious thing. I was meant to spend the night in the police station a couple of years ago. A cop I know sees me there, takes me home immediately, fetches me early the following morning to take me to the magistrate. Why does he do that? Because I am Bra Mtutu. I buy him a beer when I see him at the shebeen. He has no dignity.'

*

Luyanda Msumza, you will recall from the previous chapter, left Ginsberg and went into exile in early 1978 to join the PAC, leaving behind his father, the gruff, patriarchal CID man.

Luyanda returned to Ginsberg in 1990 after an absence of nearly 13 years. His dad was retired now; as a policeman, he was a figure of the past. I ask Luyanda how the standing of black policemen in Ginsberg has changed since he was there last.

'Well,' he replies, 'compare my position to theirs. We were both coming home to Ginsberg after an absence. I could explain mine. I could tell the young ones who'd never met me where I had been and why I had gone away. The police who came back, they could not do that. They had things to explain to their neighbours, but they could not explain. They were ashamed. They had lost their dignity.

'The uprisings robbed the cops of their standing, and they have not recovered it to this day. You don't see the pride you saw in my dad's day. Yes, the respect people had for him was mingled with fear, but he was held in higher regard. It is sad. The black community needs the sort of service my dad gave and we have trashed it. We were not wrong to do so, but the price was high.

'The teachers and the cops of the old days were very respected people. After all, it was black kids who the teachers were educating. Some of us are very grateful to those teachers today. Without them, we would be nothing. It was not so different with the cops. It was black people my father was helping to solve disputes, crime in the black community he was called upon to deal with. When the system started collapsing, these things collapsed with it.'

*

When the police's new political bosses came to power in 1994, they gave themselves the tasks of stabilising the organisation, knitting its allegiances to the new democratic order, and winning legitimacy for it in black communities. Perhaps the last of these tasks was misconceived. Perhaps the proper aim ought not to have been to win the police legitimacy, but authority.

At bottom, the liberation movement that came to power in 1994 understood South Africa's urban black population to be politically homogenous in the last instance, a people united by a common experience of oppression. It understood the violence that had wounded township life in the dying years of apartheid to be circumstantial, and thus superficial. And so its conception of urban South Africa was of a people quite ready to give its consent to being policed. The government's task, it believed, was to present people with a police force in which they could trust and believe.

Yet the violence was neither circumstantial nor superficial. It was an inflammation of the very tissue of urban South African life, a tissue in which protection and coercion coalesced around allegiances and markets. To get South Africa to give its consent to being policed would require breaking down a generations-old architecture of security and protection. It would require bringing a body with unprecedented authority into township life, a body elevated above existing security markets and thus able to break their logic. A body without that authority would simply have to negotiate its way into existing markets, becoming yet another player among many.

Constables K and N, who were punished by civilians for policing their jurisdiction in Randfontein on a Saturday night; Dan Sibanda and Joe Mukwali of Alexandra, who must patrol their neighbourhood without quite knowing whether the police are on their side: indeed, all the characters in these pages are inheritors of a police force that has never risen above the systems of protection and violence formed under white rule.

It was never going to be an easy task. But it was made more difficult by the fact that the liberation movement that took office in 1994 misread its people. It believed that its people had long ago given their consent to being policed, that it was just a question of delivering a benign police force.

If the task had been properly recognised, how would the government have set about accomplishing it? I have only ever come across one answer to this question: it is in a book by Antony Altbeker called *A Country At War with Itself* published in 2007. Altbeker argues that the police's new political bosses in 1994 should have understood their most urgent task as that of rebuilding the detective service. If the police service did stand a chance of elevating itself above existing security markets, it was by doing well what states alone can do: detecting violent crime with competence and impartiality and seeing to the prosecution of offenders. That is, after all, precisely what had been missing from generations of township life.

Constable T and the impossible suburbs

Here and now in the late 2007, some might be tempted to bring back to life the ghosts of cops like Detective Msumza. If he and his sort were the last black police officers in living memory to wield authority in their neighbourhoods, ought we not consider the prospect of resurrecting them?

It cannot be done. For better or worse, the townships will never see his kind again, and the central reasons have little to do with policing and security. The Detective Msumzas of the world have been lost in the shifting sands of South African class formation.

In his day, Detective Msumza was considered a township petit bourgeois, a man elevated by his occupation. Today's police are also considered petit bourgeois, but what that has come to mean has changed profoundly. While Detective Msumza's place on the lowest rungs of the middle class lent him dignity, today's young cops carry the burden of being considered unjustly lucky, and scavengers and thieves to boot.

*

In mid-2007, I spend a week on the job with Constable T. He is 28 years old, has been a policeman for two-and-a-half years, and is already a sector manager in a large township on Johannesburg's East Rand.

He asks me on our first day together for how many years I studied at university. He is extraordinarily wiry and thin, his speech rapid and urgent; my first impression is of a man who consumes the energy he imbibes too fast, a man who is neurotically busy.

I tell him I was at university for a decade. He whistles through his teeth.

'Is your mother still alive?' he asks.

'Yes.'

'If I was her, I would wrap you up in cotton wool. I would give you antibiotics every day just in case you get sick. I would never allow you to go anywhere it is dangerous. She should take out insurance on you. You are more valuable than a room full of diamonds.'

The sector Constable T manages is home to some 70 000 people. It includes blocks of formal houses occupied by fourth-generation families, shack settlements of Zimbabweans, Mozambicans and an assortment of recently urbanised South Africans, as well as a single-sex hostel.

When I tell him on the third day of our acquaintance that I wish to write about the forces thrusting him towards a life of corruption, we immediately agree that I will not mention the police station at which he works, the suburb in which he lives, or the place he was born.

Constable T grew up in one of the former homelands scattered across the northern reaches of South Africa. The first of his mother's seven children, he knew poverty as well as anybody. The family's primary source of income was its matriarch's wage; she was a domestic worker for a white family in one of those one-horse rural towns where the maids and gardeners earn next to nothing.

It is not unusual for a breadwinner in a very large, very poor family to back one of her children as a punter does a horse. There is enough money for one kid to be seen through a decent education, and so the matriarch chooses early which of her kids is to be the one. Constable T was the horse his mother backed, and so when he arrived in Johannesburg seven years ago at the age of 21, he was armed with a high school education and a place at a technikon. He was to study production management.

Before the year was over, he had been expelled from the institution for failing to pay his second instalment of fees.

'There was actually never going to be enough money for me to pay for the course and to live,' he tells me. 'My family was dreaming.

'So I signed up at one of these agencies that provide temporary work. I got to know every part of this city. I handed out adverts at traffic intersections in Sandton, in Benoni, in Boksburg, and in the south; I can still taste the exhaust fumes when I swallow. In Midrand, I handed out boxes of free condoms to the people in their cars. I once got work at KFC. I wore a chicken suit; my job was to be this very big chicken who plays with the children while their parents are eating lunch. That was my favourite; playing with children is not really work, is it?'

Then Constable T got what he considers the most serious job of his pre-police era: he worked for some months at Johannesburg's rail container depot, the central point at which sea freight from Durban is sorted and distributed throughout Johannesburg.

'It was the first job in which I saw a system, and I got to understand how the system works,' he says. 'I got very interested in the work of the Customs people. I thought to myself: if this were my job, I would be quite happy.'

Twice, Constable T applied to join the police academy.

'I don't know what happened to my application first time,' he tells me. 'On the forms, the only contact I left was my cellphone number, and I lost my cellphone the week after I submitted my application. So who knows? Maybe they phoned me.'

They did phone him on his second attempt, and the man who until now had on some days dressed up as a chicken, and on others stood in the traffic handing out adverts, was catapulted into an aspirant middle class.

That is the story of all the young constables I meet. On an evening in August 2007, I find myself in a police van on the East Rand with five constables, the most experienced of whom has been in the police less than four years. I ask each of them in turn about his respective employment history.

'I was the assistant at a vet's surgery,' one of them says. 'My job was to muzzle the dogs who looked like they were going to bite, and to drag them to the operating room.'

'I was a security guard at Sappi's headquarters in Braamfontein,' says another. 'I had the graveyard shift:

10 pm to 6 am. Only in my second month did I meet someone who worked in the building.'

'My cousin is a mechanic,' says a third. 'I helped him fix cars.'

When the police advertise for applications to the academy, they receive thousands upon thousands of applications. The majority of those who apply are qualified for the job: they have a Senior Certificate, a driving licence, are generally in their mid- to late twenties, and have some work and life experience.

And yet only a sliver of them get the phone call that Constable T got.

I think of this vast crowd of applicants as the slot-machine players at a giant casino. Each puts his one-rand coin into the slot, in the hope that a new life will tumble into the tray. For the lucky few, that is what happens.

*

What it meant to be a petit bourgeois in Detective Msumza's day had very little to do with his salary or his assets.

'People in Ginsberg called us a middle-class family,' Luyanda tells me, 'but really, we lived in a two-roomed house with an outside toilet like everyone else. Maybe we were the first to get a radio. What people meant by middle-class was professional. My father wore a collar and tie. He worked with pen and paper. That made him a respected person.'

The point is made with admirable clarity by Adam Ashforth in his ethnography of Soweto, *Witchcraft,*

Violence, and Democracy in South Africa. What it meant to be middle class, he points out, could only be very limited when black people had no choice but to rent a township house from the state, no choice but to send their children to Bantu-education schools. Ashforth speaks of an elderly Soweto couple he knows, the Mfetes, considered back in the 1960s to be middle class.

'None of their friends and family had a better house,' Ashforth writes. 'None lived in anything notably worse ... More money, were it to have been available to the Mfetes in the 1960s, would have had only limited utility after the rent was paid and the groceries bought. The money they had was sufficient to mark a degree of social distinction by expenditure on clothes ..., furniture (though not electrical appliances, as there was no electricity available), and, most important, a car.'

How foreign this must sound to a young middle-class aspirant today. Now, to enter the petit bourgeois, and to make the sort of investments that will keep your children there, is dearer than the Mfetes and Msumzas of that old world could have imagined. To ensure that your children attend a good school, you must buy a house in the suburbs. You have no reserves of cash, no investments, and so your entire house is bought with borrowed cash; the Reserve Bank governor's quarterly decisions on interest rates, which once meant so little that you were barely aware of them, can now destroy your precarious monthly budget overnight.

And then there is the school itself; the decent ones cost money. There is private healthcare, increasingly obligatory to any suburban family. One cannot do without a car now that each family member must commute daily through suburban sprawl, and that, too, must be bought on credit. And when your children finish school, they must without question have the means to go to university.

Constable T has a wife and a four-year-old daughter. He still lives in the township, but he has bought a house for R400 000 in a suburb 25 kilometres away. He put down a deposit of R20 000; the house is almost all the bank's. In two years time, his daughter will enrol at a respectable suburban school. It isn't cheap. Constable T earns about R80 000 a year. His wife brings in another R50 000. The accumulated cost of their monthly bond, rates and taxes, utilities bills, their daughter's school fees, and the payments on the family car, takes nearly two-thirds of their monthly income.

Then there is Constable T's family up north. His mother has not said so openly, but he can sense that she is waiting for the day he tells her she can give up domestic work, that he will look after her. As for his six younger siblings, the four who have finished school are all unemployed. They know that he is moving into a house in the suburbs, something quite beyond the bounds of experience. That he will be a benefactor is so deeply expected it has never been explicitly mentioned.

'Only you whites are able to keep the money you earn,' Constable T remarks thinly. 'You can keep your money when you come from money, not when you come from poverty.'

Constable T's story is typical. On the August evening when I drive through the East Rand with the five young constables, each tells me that he has bought property in the suburbs since joining the police, and each complains that he is drowning in debt. Nor is it just the younger police officers. On the Saturday night I spent with Captain R in Alex, he was doing overtime, as he does every weekend he possibly can, to contribute to the bond on his house in the suburbs, to his daughter's education, and to his patriarch's luxury: a subscription to satellite TV.

An R80 000 a year job is not a ticket into the middle class. And yet when one joins the police, the expectation that one will dive into a suburban existence is formidable. Indeed, it is more than formidable, more than simply an expectation: it is irresistible. It seems to be as much a part of the job as one's service pistol and one's cap.

This story of the impossible journey to the suburbs is etched into the very core of what it means to be a young police officer in today's townships. When a township resident passes a cop on the streets, she knows there is a strong possibility that he has recently moved to the suburbs, that he cannot afford to be there, and that he must use the licence his work affords him to make more money. He is assumed to be a scavenger. It is in the nature of his position, an essential part of him.

*

On our second day together, Constable T tells me that his and his wife's combined income will not sustain their move to the suburbs.

'Why then are you moving?' I ask. 'Why try to live beyond your means?'

He snaps his head sideways to glare at me, taking his

eye off the road. There is anger in his look, but it quickly evaporates; he forgives me, for I am ignorant.

The answer borne by his silence is this: they have made me a policeman, and thus a middle-class person, but they do not give me the salary for it. Your question, he is telling me, is inside out.

'How are you going to fund your move?' I ask.

In the back of the car is a middle-aged civilian. His name is Bra Jack, and he is the chairperson of the police-civilian crime forum in Constable T's sector.

'He has three options,' Bra Jack chips in. 'To make business, to make cho-cho, or to find a better job.'

'Cho-cho' is township slang for graft: for the traffic cop who pockets a hundred bucks and ignores that your car has no registration, for the detective who takes a wad of notes to work hard on your case.

Constable T smiles at the provocation, politely, but without pleasure. 'I have applied for another job,' he says. 'At Customs. If I get it I will go back to the container depot, but this time in a good job. They will put me on a training course, I will become a Customs Officer, and I will earn maybe R100 000 a year, maybe even more. I think I have a good chance. They know me at the depot. I was the best temporary worker they had ever had. I understood the system. I understood how to fix it when something went wrong.'

'And if you are stuck in the police?'

He shifts his body weight, hunches into the steering wheel, and shrugs his narrow shoulders. His eyes scan the houses and shacks and the rows of small shops outside our window, as if to say that his income will come from the scene in front of us.

'Maybe I will run a taxi,' he says. 'After you have paid your driver and your petrol and your maintenance, you make maybe R5 000 a month pure profit.'

'It will have to be in his wife's name,' Bra Jack remarks. 'Otherwise the police will fire him.'

'How will you raise the capital?' I ask.

'That's the thing,' he replies, shrugging again, this time with resignation.

'A decent minibus will cost what,' I continue, 'R80 000? That's 15 months' profit just to pay it back.'

We drive in silence for a while.

'He will find capital on very favourable terms,' Bra Jack says from the back. 'He is lucky. There is a big taxi rank in his sector.'

I glance at Constable T. He stares straight ahead. I understand what Bra Jack has just said, but it is somehow unsatisfying to hear it couched in such delicate innuendo. Stated in these terms, it is so ephemeral, it may vanish into the air. But his meaning is quite clear; T is lucky that there is a big rank in his sector, for it means that there are wealthy taxi operators invested in this local branch of the industry, and they will be very pleased to have the police sector manager up to his eyeballs in their debt. If Constable T starts asking around, there will be no shortage of people willing to help him buy a taxi. That is what Bra Jack is saying.

Constable T is going to the suburbs and will do what he must to stay there. That much is a *fait accompli*. Which means that he must either stop being a policeman or become a compromised policeman. He must either get a better paying job or get financial help from vested interests in his jurisdiction. Or he must do 'cho-cho', as Bra Jack pointed out: he must seek graft where he can.

Bra Jack is sharp-tongued: the elderly have licence to address the young with a dollop of sarcasm. But what strikes me is his resignation. That he is with us in the car is no doubt a measure of his civic duty; chairing the crime forum is an investment in his neighbourhood. That the police officer with whom he must work will make business or make cho-cho is something to which he is irritably reconciled. Constable T, he seems to be saying, is as good as it gets.

*

On another day in his patrol van, Constable T is talking about the difficulties his youth presents for his work.

'I am the sector manager, right. That means people like Bra Jack come to me and say A and B and C is wrong in this sector. Please fix it. And this means going to the head of crime prevention, or to somebody, and saying: "Please do A, B and C in my sector." And he says to me: "You joined the force yesterday. Your voice is not very loud yet. I can't hear you. Come back to me in a few years."

'And then,' he continues, 'there are the criminals in my sector. They drive past me in their BMWs and look at me, and the way they look at me is saying: "You are little. You are nothing. You cannot hurt us. You can only get hurt."'

'So then where do you get the authority to do your job?' I ask.

'You develop ways and means,' he replies. 'I have

been here two years and I have not yet turned a blind eye to anything in my sector. If somebody here must be arrested, I arrange for other people to do the arrest. That way, I can still do my job here with freedom.'

That seems a pretty tenuous proposition to me. From my vantage point, the forces conspiring to make of Constable T a policeman who is less than honest are ineluctable. Through the eyes of the civilians in his sector, he is a young man whose uniform confers upon him the most fragile and unconvincing authority. To be sector manager here, he must seek a modus vivendi with people far more powerful and resourceful than he is. By his own admission, when he makes an arrest he must hide.

More than that, he is living beyond the means his salary provides him, and although not everyone acquainted with him will know the details, they probably take his indebtedness as a given. It is written into his young face, his accent, the uncertainty in his gait, into the fact that he is a young man from the homelands lucky enough to have become a policeman.

He is thus a figure of weakness and of neediness, and the combination is pretty lethal. It is expected that he will take the crumbs offered to him in exchange for not minding his business, the business of being a policeman. The utter contempt Mtutuzela Matshoba expresses for cops is echoed in interview after interview. That township police are too weak and too corrupt to be effective is a presumption, a fact of life that inspires bitterness.

*

To which I should hasten to add a very important qualification. 'Township police' is not quite accurate. 'Black, male township police' hits the mark.

Mtutu is echoing the vast majority of township residents I interview when he points this out.

'I'll be the first to admit that it is a prejudice,' he remarks. 'There is a perception that it is better to deal with a white policeman than a black one, and that among the blacks, it is better to deal with a woman than a man. The white one will not be corrupt; he is just a professional. The woman: she is interested in dealing with your case professionally and seriously. It is the black male policemen who are scorned. We believe that they are all disconsolate, they are all alcoholics, they would do anything for a few beers.'

When Mtutu says 'we', he is explaining things to an outsider, as a Muslim would say, 'We do this and we don't do that' when explaining the laws of halal to a Christian. This is a matter between township men, he seems to be saying.

I point this out, but can only speculate what it is about. Perhaps it is this. Those treading the paths to the suburbs are becoming family patriarchs. With their new houses and cars, their satellite TVs, their broods of kids on medical aids and extra maths lessons; they are the little emperors of a brand new domain.

And yet, on the journey there, they stand on other men's heads. In the townships they are leaving behind, they do cho-cho and make business, as if they are exacting a private rent from other men in order to fund their own dream. They denigrate public township services to advance their personal campaigns. And so they are spitting on the men they leave behind.

The end of apartheid has created a new class of black patriarch, the suburban man, to which so many aspire. But the available spaces are very few, the applications vastly oversubscribed, and those who get there must scramble and kick. They must become like Constable T and contemplate licking the arse of the local taxi boss, or doing cho-cho on a street corner. Perhaps that is the story township men tell themselves about black male police officers.

*

The week in which I accompany Constable T on the job is a raucous one in South African politics. President Thabo Mbeki fires his Deputy Health Minister, Nozizwe Madlala-Routledge, who in AIDS circles is lionised for having stood up to her seniors and delivered a long overdue programme to tackle the epidemic. The following week, the *Sunday Times*, increasingly tetchy with Mbeki and a great supporter of Madlala-Routledge, retaliates. On its front page it hoists a story it has been hoarding for months: the Health Minister, Manto Tshabalala-Msimang, is an alcoholic.

Constable T asks me what I think. I tell him that the minister is unfit for office.

'No,' he replies. 'What happened was this. Mbeki fired Nozizwe because Manto hated her. Nozizwe was very angry. She took revenge. She went to the *Sunday Times*.'

'What's that got to do with whether Manto is fit for office?' I ask.

'Everyone drinks,' he says. 'Almost everyone drinks too much. If we were to say you can't hold office if you drink, we would have to hand the country over to the five-year-olds.'

Bra Jack is in the backseat again. He chooses to say nothing. I, too, am quiet. In the silence, Constable T broods.

When he finally speaks, what comes from his mouth is a sudden storm of thoughts and associations.

'I will never vote for the ANC again in my life,' he begins. 'They are corrupt from top to bottom. Next election I will vote DA.'

I laugh. Bra Jack snorts. 'You think the whites have been vaccinated against corruption?' he snaps irritably.

'No. But their style of corruption is much better. The blacks want a million rand and they want it now, and they think they can take it without anyone looking. That is stupid. Of course you are going to get caught. And when they are caught, they cry like babies.'

'And the whites?' asks Bra Jack.

'They are patient. They wait. They scheme behind the scenes while everything looks very professional up at the front. Business as usual. But when they retire: yo! You discover that they have become filthy rich.'

Again, Bra Jack and I are both silent.

'Look at how the cops are treated now compared to the old days,' Constable T finally continues, as if this is the natural place for the conversation to have progressed. 'People feared the white cops. You would watch them drive past and your knees would tremble. You'd thank God they drove on and didn't stop. Us, people swear at us. They treat us like shit. I do not like being treated like that every day of my life. I want to instil fear in the criminals. I want their knees to shake when they hear I am coming.'

'You are a bloody fool,' Bra Jack says evenly. 'Those cops were hated as much as they were feared. When they turned around, somebody would put a knife in their backs. Is that what you want: for people to fear you and then kill you?'

'Bra Jack,' Constable T says plaintively. 'You are not listening. I do not want to be hated. I do not want to be evil. I want to be respected.'

A call comes over the radio, Constable T must attend to it, and his complaints and fantasies are swallowed up by the business of a township afternoon. A group of women from a shack settlement meet with Constable T to complain of muggings in the portable toilets at the edge of their settlement. A delegation from the women's hostel says that the rule that expels the male children of hostel residents when they turn 12 is not being enforced, that there are teenaged boys in the hostel, and they are making trouble.

That night, as I drive home from the East Rand, I think of the whites Constable T admires and envies, and I am not surprised. Over the last half-century, the vast majority of white South Africans have been given, as their birthright, a suburban existence, or one of its rural equivalents. With democracy, Constable T and thousands upon thousands of young black men and women aspire to infiltrate this white citadel and make lives there. But his journey is scrappy and tainted and difficult, as it has been for every first generation of petit bourgeoisies across the world. He feels fouled and humiliated. He looks at the white politicians and policemen and he sees ease: the ease of having done all of this before, of having been born to it. And he wishes from the very depths of his being that, like the whites, he was already there. He wonders whether it is even possible to get there if you have not already arrived. For from his vantage point, he is caught in the teeth of a cruel paradox: the path to respectability is so horribly unrespectable.

Grilled chicken, boiled rice

On a mid-winter's Friday night I accompany a uniformed patrol at Constable T's police station in the East Rand. I am not with Constable T tonight, but with a much older man, Inspector R, and his younger partner. Inspector R is humourless and visibly unhappy. He is clearly far more than indifferent to his job; it gets under his skin – he hates it. From midnight, when the streets begin to empty, he searches incessantly for young lovers walking home. When he finds them he drives into their paths, shines his brights in their faces, puts his head out of the window, and curses them for being out now, when the streets are dangerous. He seems to take it all personally, as if they have chosen to put themselves at risk in order to make him miserable. His grumpiness is infectious; I withdraw into myself to escape it, and find my mind wandering to other police stations and other police officers.

Just before 1 am, a dispatch comes over the radio. A person has been shot. The address is a small informal settlement in the middle of our jurisdiction.

Inspector R takes us up a steep, rutted road that is badly in need of repair. It widens after a short distance and opens onto a dirt courtyard behind a tall block of flats. The block overlooks a small, square squatter settlement. It is crammed tightly into its space, the shacks on its perimeter almost rubbing against the walls of the apartment block.

About half a dozen silent people are waiting for us. They lead us into the slim path between the settlement and the building, and as we enter it we rearrange ourselves into single file. Ahead of us in the path, I can make out the dark shape of a body. It lies on its back, its head pointing at the shack settlement, one of its feet at the block of flats. The other foot has twisted around its ankle and faces upwards and inwards, towards the body's crotch. Inspector R shines his flashlight at it, and what we see is a young man, 30 perhaps, or a little more, his eyes round and wide open, his flabby belly exposed on this cold night. For a moment, I am certain that his eyes are taking us in, registering our presence. I must blink my own eyes several times before the assumptions in my head recalibrate, and it becomes clear that his stare is blank.

A man steps out of the gathering and introduces himself to Inspector R as the dead one's brother. Inspector R takes out his notebook and asks quietly for the name of the deceased.

'Ephraim Banda,' the brother says.

'What?'

'Ugama lakhe uEphraim.'

The brother's story is this. He, Ephraim, Ephraim's wife, and one or two others were drinking in their shack. Shortly after 12.30 am, they ran out of beer. There is a shebeen less than half a kilometre away. Ephraim and his brother went to buy some beer. Less than a minute into their journey, two men stopped them. One pushed the barrel of a pistol into Ephraim's stomach. They wanted money.

The brother was the one carrying the cash. He put his hand in his pocket and took out four hundred-rand notes.

'And you?' the man with the gun asked Ephraim.

'I have no money,' Ephraim replied.

To which Ephraim's assailant replied by lifting the barrel of his gun from stomach to heart and pulling the trigger.

I search for Ephraim's wife. At the end of the narrow passageway in which we stand, lit dimly by a distant streetlight, a woman sits on a rock. Her palm cups her chin, and her head is absolutely still. She will stay in that position, barely moving, until we leave at dawn.

Those gathered around the body are to stand watching in silence throughout the night as the rituals of murder-scene procedure are performed before them. First, the paramedics arrive, two young whites, a man and a woman. The man crouches over the body for a short while, makes a cellphone call to ask a superior for permission to write a declaration of death, nods cursorily at Inspector R, then leaves. He is followed by a young, slim, handsome man in Diesel jeans and good shoes who busies himself with tripods and structures of metal poles and bolts and tools until he has erected a medley of studio lights around the body. At the flick of a switch, a new scene is conjured, a harsh-lined square of naked, violent light suddenly exploding onto the edge of this dark settlement, making stark shadows from everything that protrudes, even the coiled strands of hair that circle the dead man's bellybutton. The young man takes ten, perhaps twelve pictures, then turns out his lights and dismantles his immense scaffolding;

in minutes, it is all tucked into a medium-sized canvas bag.

After he has gone, leaving the body to Inspector R's pale torchlight, a detective arrives, a large man in a thick woollen cap from beneath which several folds of skin cascade down the back of his neck. He takes out a fresh docket, crouches next to the body, a clipboard against his thigh, and writes for what seems an interminable time. Before he arrived, Inspector R had found a spent cartridge on the ground next to the body. Now he gives it to the detective, who seals it in a plastic bag.

Meanwhile, the radio in Inspector R's van has brought news. In the hour since we have arrived at this scene, three more murders have been reported across the township. This police station only records some 70 murders a year, I find myself thinking, less than one-and-a-half a weekend; this is an unusually foul Friday night.

As the night grows colder, Inspector R grows more agitated. From his distracted pacing, and from the deep creases that have appeared in his frown, it is clear that this scene, and the other death scenes mushrooming around the township, have rattled him. At about 3 am he walks up to the people huddled together outside the police line he has erected next to the body. They have barely exchanged a word with one another all night. They are in silent shock, neighbours of the dead man, people who knew him well, who perhaps saw him and greeted him earlier this evening.

'Look at what has happened,' Inspector R says to them. 'It is because you do not report illegal firearms. This is a small squatter camp. You know. You know who brings guns here, and you do not come to us. And now,' he points his thumb over his shoulder at the body, 'now here is your Ephraim.'

As I listen, I grow angry with him. His comment is grimly wounding. It seems that he cannot help but scold. Whether it is lovers walking home, or the neighbours of a murdered man gathering around a corpse, Inspector R needs to tell civilians that it is all their fault. He has no shame, I think to myself.

But in time to come, as I follow the investigation, I will reflect often on what Inspector R has said, and I will see that he has come to the very heart of the matter.

The mortuary van arrives shortly before dawn, the body is transferred to a stretcher, and the van's back doors are opened wide. The gathering moves like a stodgy fluid, its oneness the same, its shape reconfigured, to the mouth of the van.

'There is still space for this one,' the van driver says, peering inside. 'For the last one we must still fetch, no.'

I move to the open door. Stacked together on a wide tray are two of tonight's four murder victims, both young and black, their heads, both of them, the shiny, glistening bald of the just-shaved. Their eyes are tightly shut, as if they are sleeping desperately, and I find myself thinking that each of these men has been wandering this world for 20 or 30 years, a long time by most measures, an infinite accumulation of thoughts thought, emotions felt, meals eaten, and deeds done, that tonight they have ceased, and that their 20 or 30 years will remain a part of this world for as long as those they knew remember them.

Ephraim joins them on the tray, and there they are,

three young black men in a row, and I am hurt and upset on Inspector R's behalf. For the fact that nobody came forward to tell the police about an illegal gun, and that these corpses are all of them black men who are young, and that this gathering here at the mouth of the van is as resigned as it is shocked, all of these things must leave a cold chill in those who will still people this township tomorrow, for they are surely stalked by a foulness too icy to ignore.

*

The investigating officer is Inspector H. He wears a thin cotton shirt ironed with scrupulous care. His suit is light grey and elegantly tailored to his body, which is patently very strong, his arms wide and thick, his chest a solid barrel. He wears a spirited rainbow tie attached to his shirt with a silver tie clip. He is positively old-fashioned. I tell him that his sartorial taste is extremely unusual for, and admirable in, a policeman. He replies with a vociferous nod and an open-palmed wave that says: say no more.

We meet ten days after the murder. It has taken some trouble to get to spend time with Inspector H. When I told the station commander that I wanted to follow a specific detective on a specific murder, the scene of which I had attended myself, he concealed his displeasure behind warm eyes and skilled banter. He said yes, but meant no, and I had had to better his slippery charm with stubbornness.

'A squatter camp murder,' Inspector H says, and he shakes his head. 'My two kinds of murders are squatter camp murders and shebeen murders. The shebeen ones are much easier. It is a fight over a girl. Someone loses his head. A person who has lost his head makes my job not so difficult, not usually.

'But the squatter camp murders: no. What happens is this. The tsotsis want to take money from the shack-dwellers. They kick their doors open in the night. Or they hold them up when they go outside to use the toilet. If the tsotsi doesn't know the victim, it is fine. He will take the money and leave. But if the victim knows him, then he wants to kill the victim because he doesn't want a witness. So he'll take the money outside the toilet and then shoot. Or, he'll walk up to a shack, spray bullets through the walls, and then walk in and take the money.'

'How many squatter camp murders get solved?'

'Very few, very few. The murder will go to inquest, but it will remain unsolved. The squatter people are too scared to talk. And in cases where they do talk, it is not so easy to find the culprit. To live in a squatter camp is very temporary. You kill someone, you leave the next day. You move to another squatter camp. Orange Farm, maybe, or you go to Randburg. You sell your shack for R3 000 and you disappear.'

'Why do people not talk?' I ask. 'This station is divided into sectors. Each sector has its own manager whose job is to get to know civic leaders, to know squatter camps like the one where this murder happened. Are there not people in the camp who will talk to the police?'

He shrugs. 'Some people will talk only to a uniform, some only to a detective. Some only to this particular policeman, some only to that one. Most will not speak.'

'And this murder?' I ask. 'You have an eyewitness.

He was standing maybe a metre away from the one who pulled the trigger.'

'Ja,' Inspector H replies, nodding vigorously in the same way he did when I complimented his clothes. 'I went over there to speak to him last week. He said he suspected certain people, but that they no longer live in the squatter camp. I told him he must point out the shack they lived in, and speak to the people who live there now to see if we can trace the ones who lived there before, and maybe raid them and find the weapon. We have a cartridge from the scene, so we can match it to the gun.'

'And?'

'When I said that, he said he was not so sure anymore who he suspected.'

'So what's next?'

'At some point, I go back.'

'When?'

He turns to his open diary. It is large and thick-paged, with a hard, heavy back. He leafs through it noisily. He is impeccably solicitous, quick to respond with eagerness to any question or request.

'Tomorrow I am going to the Eastern Cape to find a suspect,' he says, his eyes on the diary, his tongue moistening his page-turning finger. 'I should be back Thursday. Then there is another urgent case to work on. Friday, I must find five witnesses to give them their subpoenas. By the time that is finished, it will be after lunchtime.' He licks his finger again and pages some more. 'Next Monday, I'd say. Yes, next Monday I will go back.'

'May I join you?'

'Of course.'

*

I leave Inspector H's office in a state of indignation. There are not so many murders in this township, I think to myself, and not so few policemen, that a killing should be treated like an old woman who has lost her cat.

What strikes me above all is this: it seems from the little Inspector H has said that the residents of the informal settlement know who the murderers are. They left an eyewitness who recognised them. They have, until recently, been neighbours of the man they have killed. They have slain him over nothing, over the pique inspired by a robbery victim with the audacity to have no money in his pockets. And still, people choose to say nothing to the police.

A murder is committed, an agent of the state arrives on the scene to find the culprit and bring him to justice, and nobody will talk to him. The question is why. Is it that they expect incompetence, and worry that a poorly conducted investigation will create more danger for them; does Inspector H's evident half-heartedness spell trouble?

At about the same time as I watch the investigation of this murder, an elderly couple I know is held up at gunpoint in their home in the northern suburbs of Johannesburg. They suspect the involvement of a woman they know well, an employee who comes to clean their house twice a week. They meet with the investigating officer and are impressed by his apparent industriousness, by his unwavering concern. And they think to themselves, is this evidence of competence? Can we trust this man enough to confide in him, to pour out our suspicions about the woman who cleans our home? Or will he merely start a fire in our lives, instilling an urge for revenge in this stranger who comes to our house? They are divided between the urge to tell him all and place the whole matter in his safe hands, and the impulse to say nothing and smile sweetly at the woman who comes to their house in the hope she has what she wants and will not do it again.

Is this what is happening when the residents of the informal settlement seal their lips at the sight of Inspector H? Or is it something else entirely? Perhaps it is not their worry about his competence, but a concern that they do not know who he is.

*

I consult with Constable T. Since I told him that I want to use him as the centrepiece in a chapter about corruption, he has become a constant confider, as if I am a Catholic priest in a confessional. The murder was not committed in his sector, but it was in his station's jurisdiction.

'What do you do when there is a murder in your sector?' I ask him.

'I attend the scene. I try to find the cause, see if there are suspects, pass on information to the investigating officer.'

I give him a brief description of the murder scene I attended. I tell him that nobody has come forward to give information, and that Inspector H appears to have conducted only one interview.

'It was 1 am,' Constable T says. 'It was dark. Nobody saw.'

'No,' I say. 'People are scared to come forward. The eyewitness himself, who watched his own brother being killed, will not come forward.'

'Yes,' he replies. 'They are scared.'

'Of what?'

'That the detective will sell the information to the culprit and they will be in trouble. People do not talk to just any cop. They talk to a cop they trust.'

'So how do you manage to do your work if only a handful of people will talk to you?'

'I am a sector manager. If I knock on a door in my sector, or if I walk up to somebody in his shop, or on the street, he will not talk to me. It is just a waste of my time. My most valuable weapon is the evenings I spend in the sector crime forum. It is lively. There are about 35 people who attend regularly. They all trust me. So if there is information I need in my sector, I spread word among these people, to ask if anyone knows the person I need information from. If someone knows the one I am looking for, that is good. They will speak to that one, and then he or she will speak to me.'

'So in this case,' I say, 'perhaps the people in the squatter camp who know who the killer is simply don't know a single policeman well enough to trust?'

'The only people who talk to a detective,' Constable T replies, 'are his informers. Nobody else will talk to a detective.'

If that is true, the consequences are very far-reaching indeed. If enough people begin to believe that investigating officers are shabby entrepreneurs, the belief becomes self-fulfilling. For if nobody is prepared to talk to a cop they don't know, then the only information that ever flows is exchanged for money, for allegiances, for loyalty. It is a game, every player is an informer of sorts, and who is to say who's working for whom? A murder investigator cannot be assumed to be the state's agent, here at the scene to solve a crime. One cannot be sure that the state has an agent attending to this matter at all.

I recall an interview I did with an acquaintance of Mtutuzeli Matshoba. I shall call him Bra D. He is a middle-aged career criminal. He is renowned in his neighbourhood for his ability to make any police case go away. He trades a good deal on this reputation; it accounts for a considerable portion of his income. I asked him what he thought of the quality of the detective service.

'I have one main grievance about them,' he said. 'They do not solve cases by doing scientific police work: DNA, ballistics, fingerprints, and so on and so forth. They rely too much on informers. You may have read in the papers last week about a big cash-in-transit trial. It turned out in court that more than half the people on trial were informing for various detectives. When I read that I immediately phoned an old prison mate of mine who has been busy with cash-in-transit for many years. I said, brother, I am on my hands and knees appealing to you. Get out of that game. Do something safer. Because these days, you don't know who's who or what's what. You are working for one policeman so you think you are safe, but everyone around you is working for a different policeman, and nobody is safe.'

Perhaps that is what Ephraim's friends and family suspect. That you don't know who's who or what's what, and thus nobody is safe. I find myself comparing their situation to that of Dan Sibanda and Joe Mukwali in Alex. This murder would perhaps not have happened on Dan's and Joe's watch. If somebody saw a neighbour bringing an illegal gun to the block, they probably would have informed one of them. And if the murder had happened, the Sector Four Patrol Group would have sat the eyewitness down until he told them whom he had seen.

That is the difference between a third- or fourth-generation neighbourhood, and a transitory shack settlement where people are relative strangers. In the latter, nobody is sufficiently well established to take control.

But there is a similarity between Dan and Joe's situation and that of the people Ephraim left behind. Both live in worlds where the presence of the state comes and goes like a blinking sign. And so both must invest resources in self-protection. For Dan and Joe, this means taking to the streets on Friday and Saturday nights, in the knowledge that they cannot stop this work as long as they live where they do, for to stop would be to leave themselves exposed. For those who mourn Ephraim, self-protection consists in keeping one's head down, in the hope that if one is sufficiently quiet one will remain invisible. They are like the elderly northern suburbs couple held up at gunpoint, who do and say nothing in the hope that the woman who comes to clean their home has made her peace with them.

*

The following Monday, Inspector H and I return to the scene of the murder. Where the body had lain, a man now sits on a barrel, the shadow of the apartment block sheltering his forehead from the late morning sun. He watches us park and get out of the car, and when he sees that we are heading his way, he stands up, walks briskly in the opposite direction, and disappears into the settlement's alleyways. Inspector H clucks his tongue.

'Look at how foolish the people are,' he says with ancient resignation.

It is unseasonably warm today, and Inspector H has left his jacket in the car. In his crisp white shirtsleeves, he sits on the barrel the man has vacated, and opens his notebook.

We hear a rattling of keys from a shack just a few paces away. A door opens, and an old man emerges, his thick hair a brilliant grey, the right side of his body collapsing onto the weight of a waist-high stick. He greets Inspector H formally, and says he has been waiting some days now for the police to come, that he was thinking of walking to the station itself to make a statement, but at the last moment thought better of it.

Once we are inside his shack he tells us of two Zimbabwean brothers, Steve and Saul, who lived in this settlement until recently. They were troublemakers, he said; they had guns, and they were not scared to show them. Among the people for whom they made trouble was the old man's son.

'A month ago,' he tells Inspector H, 'my son gave Steve a good hiding, right outside in front of everyone. Somebody asked this Steve if they should call the police so he can lay a charge of assault, and he said no, he will deal with it in his own way.

'The next night, I was sleeping in my shack, and there were gunshots, and the glass next to my bed shattered, and then my foot was in great pain.'

He is sitting on his bed now. He rolls up his trousers

to show us a black, protruding stain just above his ankle.

'It only scraped,' he continues. 'It is only the skin and some flesh. No bone.'

The Zimbabweans disappeared after that, the old man says. They were gone the next day, their shack sold to a Mozambican woman. But on the afternoon of the murder Steve was back; he was seen walking up and down the shack settlement. He told people he was living in Hillbrow now. He was walking with a friend of his called Chookies, who lives in a shack just three or four rows away.

At lunchtime on the day after the murder, Steve was seen again in the settlement, and was asking questions. Did something happen here last night? What happened to Ephraim Banda? He was told that Ephraim was dead, and then he disappeared again.

*

Back in the car, Inspector H tells me that Chookies and his shack are his target.

'The gun will not have left Alex,' he says. 'Steve would have left here on a Saturday afternoon or evening. He would not have carried a gun with him from Alex to Hillbrow. It is too risky.'

'Surely it isn't,' I suggest. 'He probably travelled by taxi. How many taxi passengers heading into Hillbrow are stopped and searched on a Saturday afternoon? Any?'

'No, no, no. It is too dangerous. The gun will be here. We will raid. When I can get some manpower one day we will raid Chookies' place, and we hope to find the gun, we send it to ballistics, we hope it matches the cartridge from the scene. And after we have arrested Chookies, he will take us to Steve. He will definitely. It will be to save himself from a murder charge.'

'If the gun is still there three weeks after the murder,' I say.

Inspector H squints into the sunlight and straightens his back.

'Well,' he replies gravely, 'it was an evil, evil murder, this man just going out to buy beer, and so we can hope that God will intervene on our side.'

I begin to laugh, first with restraint, then quite loudly. Inspector H stares at me a moment and giggles uncertainly. He is a man who likes to keep good form, at his most comfortable with a steady flow of solicitousness and good cheer; this unexpected laughter has unnerved him. He giggles again, with more confidence this time.

'God!' he says finally, 'Ha!' And his giggle breaks into a laugh.

*

By the time Inspector H's office is in sight it is after midday, and I offer to buy him lunch.

'Constable T and I ate at a shebeen around the corner yesterday,' I say. 'The pap and wors is delicious.'

He grows uncomfortable and changes the subject. I raise it again.

'It's in walking distance,' I say. 'It is just here.'

'I'm happy for you to buy me lunch,' he confesses finally, 'but not in the township. We must go to the suburbs, to Nandos.'

'Okay,' I reply. 'That's fine. What's wrong with the township food?'

'Too heavy. Bad for me. At Nandos there is boiled rice, not pap. And the chicken is grilled, not fried.'

'You have a cholesterol problem?'

'No, no, no.' He laughs briefly, a single 'Ha!'

'No, my cholesterol is fine. It is that I am in training. I am a swimmer.' He looks at his watch. 'I knock off at three. I am in the Virgin Active at the East Rand Mall by 3.30 pm. My training session is two-and-a-half hours.'

'Every day?'

'Six days a week. Not Sundays. The body must rest for one day.'

'You have a coach?'

'I have swimming partners. We train for the competitions together. We meet every day at the gym at 3.30 pm. Sometimes not everyone makes it. Some must work late in the afternoon. Some are lazy. I am there every day.'

At Nandos we both order grilled chicken and boiled rice. To drink, I have coffee, Inspector H a diet coke. As he eats, I watch his powerful shoulders under his thin cotton shirt, then his big, barrel chest.

He watches me watching.

'I started to swim four years ago,' he says. 'I weighed 76 kilograms. Today I am weighing 99.' He blinks momentarily at the chest I have been examining. 'This is all work.'

'Why did you start?' I ask.

He shrugs, puts a forkful of rice in his mouth and chews.

'It annoyed my wife in the beginning,' he replies. 'No

red meat. No processed starch. Nothing fried. She had to change her whole diet. But she understands now. She understands that a man like me, if he is getting no satisfaction from his job, he must find something else.'

'Is that why you started?' I ask. 'It's a kind a replacement vocation?'

He shrugs again, scoops some more rice with his fork, puts it carefully into his mouth, and chews. 'I have been an inspector 15 years,' he says finally. 'I know that promotion is impossible for me. My salary now, it is for life. My job now, it is for life. My rank now, it is for life. I do not care about the job. Nobody does. We are all in the same boat. Caring about the job is something for the station commissioner, not for us.'

I phone Inspector H a week later to ask whether he has raided Chookies' shack. He does not return my call. I phone the following week, and the week after; I get no answer to my messages.

Refuge

The great exception to the general story I have been telling is the policing of people in their own homes. Out on the street, especially where public space is crowded, South Africans have not given their full consent to being policed, not by the police force they have, at any rate. Yet behind the closed doors of the nation's private homes, demand for the presence of police is abundant, a cascade that begins to fall upon dispatch centres every Friday evening, and does not cease until late on Sunday night.

On the streets, the young uniformed officer is a skittish figure who must beware the living and tread lightly upon the dead. He steps into a house or a shack, and his fear vanishes: he is a patriarch's patriarch, commander of the elderly and the little ones alike.

From where does this authority come? Looking for its roots in the past seems counterintuitive at first blush, since the very spectacle of domestic policing is entirely new. Everyday life was not policed under white minority rule; cops did not go into black people's homes to stop a man from beating a woman, let alone to arrest him for it. And it was unusual in the extreme for women to call the police to protect them. The very notion of anybody – a neighbour, a relative, or a cop – storming a township house and dragging a man away from those he is hurting, has little pedigree.

So how are we to explain what domestic policing has become? I suspect that it is borrowed from the old after all. Cops may not have entered township homes to keep the peace in the past, but they certainly did come in the middle of the night, to search for pass offenders, mainly, and nobody who lived in a township back then has forgotten the sight of people being dragged from their homes. Here and now, watching two police officers walk into a house to which they have been called, grab a man by the scruff of the neck and throw him into the back of a van, is to recall Mtutuzeli Matshoba's blackjacks, who knocked on the door at one in the morning, sending grown men and women scurrying for their passbooks.

There is a sense in which the blackjacks of old are kept breathing both by township men and by township women. Men because it remains in their pedigree to submit to uniformed officers who enter their homes; women because they have sensed this atavistic fear in their men and have fashioned it into an ally. They have rekindled apartheid's nastiest instrument and turned it upon the intimates who hurt them.

If this is the case, there is some irony here. It is the Domestic Violence Act of 1998, among the most distinctively post-apartheid pieces of legislation written since 1994, that obliges the police to respond energetically to women's calls.

While most self-respecting police officers would deny it, they gravitate towards domestic conflict. Some of them might treat the women who come to them badly, imposing careless and destructive solutions upon the problems they are called to solve, but the domestic sphere has become their natural home. It feels good to be there. It is a refuge from the streets in which they are put upon, insulted, and threatened. It is the one sphere in which their authority is rarely questioned; they are there because their presence has been demanded. And it feels good because it is the only part of their work that allows them to express themselves as moral agents, after a fashion.

And yet they are also right when they insist that they hate it. It is hollow work, and they know it, and much of what they do in private homes is an attempt to fill up the hollowness with whatever comes to hand.

*

Recall a moment from some time back. We are in Alexandra on a Friday night with two officers, Inspector L and Sergeant Z. It is around 1.30 am, the streets are emptying, and we come across a lovely scene: two lovers are embracing under the cover provided by an enormous winter coat. Inspector L spits venom at them and chases them off the street. He is getting empty revenge for the night he has had. He and his partner have spent much of their time dodging crowds. Sometimes they failed and were forced to police before an audience: like when they encountered Mozambicans drinking beer in the street, and when they saw a young man slipping a knife into his trousers. On both occasions, they were forced to obey an opaque set of rules laid down by the people on the streets, rules designed primarily to cushion their humiliation. Just two officers on a Friday night in Alex: they were cowed.

About half an hour has passed since the lovers in the coat-tent were sent home. We are in the heart of old Alex. The crowds on the street have thinned considerably, but there are still people about. Among them we see a woman and a small child. She is in her mid-thirties, perhaps, he no more than three or four years old. She clutches his hand tightly; she is moving with purpose.

Inspector L sits up in his seat and arcs his body towards them. When he speaks, there is in his voice a hectoring rage.

'Child abuse!' he shouts. 'This is nothing but child abuse! What is she doing on the streets with such a young child at this hour?'

He rolls down his window and asks her threateningly what she is doing, where she is going. I am thinking that this is to be a replay of the scene with the lovers and the coat, only far uglier; Inspector L intends to spend what remains of his shift seeking redress for his humiliation from the good-humoured and the weak.

When the woman begins speaking it is clear that she is deeply upset. Her voice falters, and the sides of her eyes grow moist. Her lips are tense; she is making a bold effort not to burst into tears.

She tells the officers that she is fleeing her boyfriend. He began drinking at four this afternoon, she says, and he started beating her at nine. She knows from bitter experience that he will not stop until he sleeps, and she has fled. She has taken the child because a small boy ought not to be left alone with a drunk man. Better to risk leading him by the hand through the streets. She is en route to a cousin, she says, about some ten blocks from here.

Inspector L invites her into the car, and we begin to drive. He asks her for the address of the place of refuge she has chosen, and then for her boyfriend's address. He consults briefly with his partner, and then we head, not for the refuge, but for the boyfriend. Inspector L has in his sights the most legitimate object for his anger he is likely to find this evening, and he wishes to have it out.

The woman directs us to an address down near the river, walks us into an old Alex yard, and points to a shack. The sound of reggae comes from it: bass-heavy, mellow and loud. One of the inspectors knocks and a man opens the door, bringing with him a welcome gust of warm air and the sweet-sour smell of ganja. He peers at the woman, then at one officer, then the other, then at me. His face breaks into a smile. He is amused. He invites us in.

There is a bed, and a plasma TV with a gigantic screen that consumes one of the shack's four walls. There is an equally large silver stereo resting on a shelf that is suspended from the ceiling by four chains. There is also a stove, the four plates of which are all lit up; the warmth of this place sinks immediately into my cheeks and my toes. The man himself is enormous, well over six feet tall, long-limbed and broad, and he moves about in his largeness with great ease.

Inspector L begins to shout at him, but the reggae comfortably swallows his voice. Beside himself with irritation, he orders for the music to cease. The man crosses his tiny room at leisure and cuts the sound. Inspector L's bark is alone now, and it echoes jarringly through the shack. He throws scorn at the man's very being: 'A woman and child alone on these streets after midnight! What sort of man are you! How can it be safer for them out there than in here! You are disgusting!'

The woman listens quietly to Inspector L's invective and, as she does so, the expression on her face grows bolder and angrier. When Inspector L breaks for a pause she steps in and takes his place. She moves up close to the big man and begins to scream, her spit spraying the side of his cheek. He raises an eyebrow – a lazy acknowledgment that this is an unusual situation, a woman screaming at him for all her worth, and he powerless to shut her up – and then he turns away from her.

Once she has started, it is clear that she is not going to stop. A dam wall has burst, releasing more bile and distress and disappointment than she probably knew she possessed. As I watch her, the contours of the alliance that has formed between her and Inspector L dimly emerge. They are borrowing from one another, I believe; she his uniform and the threat of his gun, and he her victimhood.

Inspector L has sought this scene. He was not called here. He did not need to stop the woman on the street. And once he invited her into the car, he did not need to bring her here. He is in this shack because he needs this confrontation. What is it that he needs from it? What has brought him here?

I have no doubt that he would deny it vociferously if I put it to him, but he is here because he identifies with this woman. The entire night he has been put upon by crowds of people who can easily hurt him, crowds by whose rules he must play if he is not to be beaten, crowds that have been emasculating him here in this township for the last 15 years. He is in search of somebody big and menacing who he can shake and shout at and turn upon until he has managed to sweat some of the humiliation out of his pores.

As for the woman, she did not invite Inspector L to this shack; she has had no part in setting up this scene. But once he is here, she scents the nature of their alliance, and she warms to it. With two armed and uniformed men in the room, men who are on her side, she is suddenly drawn into this moment. For as long as their presence lasts, the big man cannot lay a finger on her. He is disarmed. And in this unlikely pause between his past and future violence, she can conjure all the pain from the depths of her being and vomit it at him, a moment of longed-for catharsis his fists would at any other time disallow.

Perhaps this is one among the many reasons South African women demand the presence of police officers in their homes. Beyond any thoughts of the long term, beyond any consideration about what will happen tomorrow and the next day, there is deep satisfaction in the catharsis, in the sudden turning of the tables, in the brief period of freedom.

*

It is easy to miss, this edgy affinity between cops and battered women, because in the ordinary course of things, many cops treat the women who seek their help like dirt.

Early on a Friday evening, I sit in the client service centre at the police station in Reiger Park, a coloured township on the East Rand. Friday night shift is a few minutes old; behind the counter that separates cops from civilians, five or six uniformed officers are signing in.

A woman walks into the room. There is blood pouring from her nose down the sides of her mouth and onto her chin. Beneath her left eye the flesh is lumpy and swollen and is beginning to darken.

'*Ek wil 'n saak maak,*' she announces to nobody in particular.

None of the officers looks up. One of them snorts derisively, his head still down.

'*Ek wil 'n saak maak,*' she repeats. '*Daai poes het my weer geslaan, en nou slaan hy ook die kind, en ek is nou moeg, enough is enough, hy gaan vanaand in die tronk slaap.*'

'Not if you don't shut up,' one of the cops grunts. 'Shut up and wait for the van.'

'*Wanneer kom die van*?' she asks.

'After you shut up,' the cop replies.

One of the officers looks up long enough to notice that the woman is dripping blood onto the bench she has found. 'Get out!' he shouts. 'You are making a mess in this charge office. Wait for the van outside!'

The woman ignores him entirely, as if he has been speaking to someone else. She has come here to get the police to arrest her man, a course of action she appears to have brooded over for some time, and she will not be moved. The cop who scolded her clucks his tongue and continues with his work. The other officers eye one another briefly, their glances raising a mutual enquiry:

should we gang up on this bleeding woman and expel her from the room? But this fleeting, incipient alliance flickers and then dies. They are too lazy, too preoccupied: there are other things to do. One of them hands her a wad of paper towels.

*

Precisely 24 hours later, I am in the same room, and the same five or six officers are preparing for a new shift. A woman walks in. Like her counterpart from yesterday, she has been beaten, her cheeks both swollen, her left earlobe caked in dried blood. She is African. She has probably come from the large informal settlement that ends just a few hundred metres from here. Her hair is long and is tied into three large bangs that point anarchically from her head in different directions.

The officer in charge of tonight's shift is a middle-aged woman, Inspector X. She looks up, takes in the woman's bruises, and lowers her head again. 'What happened, sister?' she asks.

The woman rests her elbows on the counter, sighs heavily, and begins to tell a story in a quiet, even voice. She is speaking a mixture of native-tongue Xhosa and recently learned Zulu, and I struggle to follow her. I find that my attention is drawn to the officers in the room. They are busy with their papers, their logbooks, and their equipment, but they are all listening. More than listening, they are drawn. They are sucked in. Their facial expressions, as they go about their tasks, are those of an afternoon soap opera audience. They are *involved*, and it is as clear as day that these women who come one after the other into this room are quenching a great thirst.

Like a television audience, the officers can lose themselves in the lives of others. But what marks them from those glued to the silver screen is that they are players in the dramas they watch; they intervene. And so they are free to take the mess of the rest of the their working lives, the humiliation and smallness they suffer in the streets, and to refract it through the domestic scenes in which they participate. Policing violence among intimates thus becomes a cheap form of therapy. Here, in the one domain where they are in control, they have the freedom to choose roles, and thus to win back something of what they lose among the crowds outside. They can treat women like shit and thus become the ones who treat *them* like shit out on the streets. Or they can identify with the women who call them, grab violent men by the scruffs of their necks, and thus get revenge on the proxies of the ones who bully them.

Most of all, they can write some of the rules, an unusual freedom in their work, and they do so with relish. Riding along with two officers in Kagiso on the West Rand some time in 2004, we received a domestic violence call and found our way to a small house on the outskirts of the township. The complainant was a young woman. She told the officers that her landlord, an elderly woman ironing clothes in the room next door, beat her regularly with a sjambok. The cops were suspicious of the story. They began asking the complainant questions. She had not expected this, and she did not like it. She grew insolent, accused the cops of not doing their jobs, and before she could finish reprimanding them, one officer had grabbed her roughly by one arm, another by her other arm; they threw her unceremoniously into the back of their van and drove her to the station.

'On what grounds are you arresting her?' I asked.

'I don't know,' one of the cops replied. 'Public drunkenness, maybe.'

'Why?'

'Because she has abused us. When you call the police, you have taken on a responsibility; there are rules you must obey.'

I marvel at this story every time I think of it. It is quintessentially a *domestic* policing story. Once we walk into a private home it is ours, the cops are saying, and they exercise their proprietorship with a voracious hunger. I marvel, too, at how those they police unquestioningly accept the rules they lay down. By the time I had finished my research in Kagiso I had watched cops arrest more than a dozen people for public drunkenness in their own homes. Not a soul protested. That the last of your rights is suspended when a police officer sets foot in your home is taken for granted.

*

What do cops do once they are inside private homes? How do they go about tackling the problems they have been called upon to address? This is among the most difficult questions about which to generalise. So much depends on an individual cop's quiddity, frame of mind, and, above all, on his or her gender. The Domestic Violence Act is a long and subtle piece of legislation;

it allows for sophisticated cops to give expression to their thoughtfulness, and blunt cops their bluntness.

But aside from some notable exceptions, which I will describe a little later, for most cops the question is whether to urge the complainant to give her consent to an arrest, or, on the contrary, to dissuade her from pressing charges. And once you have spent a few nights watching this pendulum swing from one alternative to the other, you understand how hopelessly inappropriate a tool this capacity to arrest is, how mesmerising and dumbing is the swinging pendulum.

'I am not a negotiating officer,' Inspector L shouted into the ear of the complainant at the first domestic violence call he received on the Friday night I spent with him in Alex. 'I do not negotiate. I arrest.'

'Why is it better to arrest than to negotiate?' I asked him afterwards.

'Because to negotiate is to postpone the problem. You talk, and then an hour later he *klaps* her, and you must come back. So you may as well arrest in round one.'

This is a constant refrain. In Reiger Park, on the weekend the two bleeding women walk into the client service centre, I spend Friday and Saturday night with two young constables, B and M. During the course of the weekend, we attend about ten domestic violence complaints. Without a single exception, the constables know each complainant well.

'But why didn't you call us earlier?' they say over and over again. 'You know we would have arrested him for you.'

'That is exactly why I didn't call you,' one of them replies.

'Well, then, this is your fault. We cannot help if you don't allow us to.'

At about one o'clock on the Sunday morning I spend with them, Constables B and M are called to a domestic violence complaint. They do not recognise the address, but when we arrive, there are two women waiting outside, one middle-aged, the other in her teens, and the officers immediately recognise them both. Constable B rolls down his window.

'Mrs Adams,' he says. 'You have fled to your other cousins this time? We have never been here before.'

She nods. 'I've had enough, Constable,' she says shakily. 'This time he was threatening my daughter. He has never done that before. I took my daughter immediately, and I phoned you.' The teenaged girl stares sulkily into the distance.

'Where is he?' Constable B asks. 'At the flat, or at his girlfriend?'

'At the flat.'

The two women climb into the back of the car, and we drive.

'When was it that we filled in the form for the protection order?' Constable B asks, the patience in his voice a little too studied, betraying a trace of condescension.

'I don't know,' Mrs Adams replies absently.

'I think it was three weeks ago,' he continues. 'Did you ever take it to the magistrate?'

She does not answer.

'No, you did not. And now look. You are chased out of your own home. If you had followed my advice you would be sleeping in your bed now, and your daughter would not have had this happen to her.'

Again, she says nothing.

We arrive at a tenement block in the middle of Reiger Park, scale several flights of a wrought-iron fire escape, and walk into a small, square flat. Mr Adams is sitting quietly on a double bed in one of the bedrooms. He is in jeans and a black leather jacket, an open can of beer resting on his thigh. He is compact, very strong and balding, and has a large moustache. He looks expressionlessly at the constables and mumbles a polite greeting.

'Are you going to come nicely, Mr Adams?' Constable M asks, as if speaking to an invalid or a very ill patient. 'Remember last time how I had to hurt you.'

Mr Adams gets up carefully from the bed and comes along, quiet and cooperative as a young lamb, treading carefully and uncertainly through his drunkenness. But when we get to the station, a small, cramped place with no facility for detainees, Constable B cuffs Mr Adams to a heavy table, and he begins to shout in protest. Cuffed to a table opposite him is a rape suspect. He was arrested earlier in the evening at an informal settlement nearby called Jerusalem. He has been badly beaten and is clearly in shock. A vigilante group meted out punishment for some time before the police were called. The group is run by an elderly woman called Sis Grace, who represents Jerusalem at the Reiger Park Community Crime Forum, and who I last saw a couple of hours back, a large sjambok in her right hand, this rape suspect kneeling at her feet and in some pain.

Mr Adams has ascertained that the rape suspect's name is Patrick.

'Patrick,' he says belligerently. 'What's up, Patrick? Patrick, what's news?'

The rape suspect does his best to pretend that he has not heard, a laughable ploy what with the violence Mr Adams's baritone is doing to this confined space.

'Patrick! Patrick, why do you *naai meisies* who don't want to be *naaied,* Patrick? Patrick, you're bleeding, Patrick! Patrick, I'm talking to you! Patrick, you must only put your knob where it's invited, Patrick. Patrick!'

Patrick has lost the pretence that he can't hear. At each mention of his name, he winces, which only spurs Mr Adams on. Constable B puts down the docket he has been working on and tells Mr Adams with some authority that he is going to kick his teeth in. Mr Adams falls silent. Only to begin a moment or two later, his attention trained now upon his wife and stepdaughter, who are sitting, out of sight, on a bench on the civilian side of the counter.

Later, Patrick and Mr Adams are taken to the Boksburg police station, where there are sufficient cells to hold them and, outside the cells, Mr Adams learns to his delight that Patrick is being arrested for the first time in his life. Quick as a flash, he dons the roll of desk sergeant, telling Patrick in a tired, bureaucratic monotone to take off shoes and belt, to find his standard-issue blanket in the top right-hand cupboard, and that the toilet in the cell does not flush, which means he should try not to shit.

'Patrick,' Constable B says, 'you are going to have to put up with this until Monday morning.'

'Monday!' Mr Adams shouts. 'I will be here much longer than that. Just yesterday morning I was in court for my last case. And on Monday morning, I was in court for the case before last. They are never going to let me out on bail, my friend. I am in big trouble this time.'

Earlier, when Patrick and Mr Adams were being led out of the Reiger Park police station, Mr Adams had caught sight of his wife and stepdaughter for the first time since he'd been cuffed to the table. He put his cuffed wrists together and pointed at his family with both of his index fingers: 'I'll be back,' he said with a flourish. 'Mark my words, I'll be back.'

That, I thought to myself, is the most perceptive comment anyone has uttered about this situation all evening.

*

Later, when Mr Adams has been locked up in the Boksburg cells for several hours, and the two constables have attended to a handful of subsequent complaints, I ask Constable B about tonight's events.

'Does it help to arrest?'

'What do you mean?' Constable B replies.

'I mean, when you arrest someone for domestic violence, and he comes back home after he has been to jail, does he beat his partner again?'

'Some yes, some no.'

'Can you tell in advance which sort of man will and which won't?'

'No.'

'Not at all? Whether he is young or old, or middle class or poor, or employed or unemployed, or a heavy drinker or not?'

'You can't tell,' Constable B says.

It strikes me as he says this that the world Constable B enters when he polices domestic violence is solipsistic. His itch to arrest has little to do with the lives of the complainant or the perpetrator; he is barely thinking about the effects his actions will have on their future. His concern is with himself; he is modulating the effects of his work upon his own psyche. For what drives him to distraction is the lack of punctuation, the perpetual sameness. He is called back to the same addresses again and again; and he is well aware that he is utterly powerless to recalibrate the relationships between the people who are fighting.

He arrests because it breaks the sameness. It kicks off a new set of procedures, delaying the start of the next cycle of sameness by some weeks, perhaps even months. It rearranges the characters into a new relationship with one another, briefly, at any rate, and thus allows him, as a spectator to the drama, to experience a new array of emotions. And so when he, together with Inspector L in Alexandra, and countless other police officers like them, tell a complainant – 'I am not a negotiating cop, I am an arresting cop' – they are pleading with the complainant to provide them with some variety.

*

Not all cops police domestic violence like Constable B and Inspector L. The exception that springs most vividly to mind is, not coincidentally, a woman. Her name is Inspector N, and I encountered her in Kagiso in 2004. She was in her early forties, was a single mother of two teenaged children, and was training to become a sangoma;

around her left wrist she wore a bracelet made from the fur of a genet: protection, she said, against those who wanted her training to falter.

I was first struck by her manner of dealing with domestic matters on a Wednesday afternoon in a small house in a respectable Kagiso suburb. The complaint was not one of domestic violence, but of housebreaking and theft. Inspector N sat at the kitchen table and listened to the complainant with great absorption. Her partner, a 26-year-old constable, wandered listlessly around the kitchen.

The complainant was a young woman. She owned this house, she said, but had moved out some months ago, leaving it to her ex-boyfriend, with whom she still shared a bed from time to time. He phoned her the other day, she continued, to say that there had been a burglary: two television sets and video recorder were gone. But no lock had been picked, she said, and no window broken. It was clearly the ex-boyfriend. He does not work, he drinks a lot, he is always short of money.

'Open a case,' Inspector N said crisply. 'Yes, it is a useless case because there is no evidence. But it is important because it puts him under pressure. Detectives come here and ask him lots of questions. He feels the pressure.'

The young woman nods, at first uncertainly, then with heightened attention.

'But there is something else you must do,' Inspector N continued. 'Tell me, how much of this house have you paid off?'

'I've been paying the bond only a year,' the complainant replied.

'Go to the bank and sell it. That is my advice. And the money you get for it, invest it in something your ex-boyfriend cannot touch. Maybe you love him, maybe you care for him, but he is a sponge. This house is just an opportunity for him to sponge off you.'

I snapped awake and stared at Inspector N. Perhaps her advice was good, perhaps poor, but the point is that she had stepped out of her skin and was examining the matter from the complainant's perspective; and she was thinking, not about the coming hours or days, but about the coming years. Having watched cop after cop turn the homes of domestic violence complainants into receptacles of their own neediness – the need to get off the streets, to exercise authority, to get the succour that the crowds outside denied them – it was refreshing and unusual to watch a cop imagine the world from the perspective of the one who had called her.

*

Two days later, I spent a Friday night shift with Inspector N and her young partner. Late in the evening, we found ourselves sitting at the dining room table of a silent double-storey house in the middle-class suburb of Riverside. The interior walls were all of unplastered facebrick, the table at which we sat a heavy, dark wood, and the place seemed hollow and cold. The woman sitting opposite us had a cut on the side of her head. She spoke in a steady monotone, her eyes wide and quite blank.

She told us that she had been married 17 years. Her three kids all lived in the house. She said that her husband beat her whenever he was having an affair, that he started a new affair once, perhaps twice a year. He had beaten her earlier that evening. He had thrown a heavy dining room chair at her. The end of a chair leg had grazed the side of her forehead. She had a protection order against him. He had violated it.

'You cannot spend the night here, sister,' Inspector N said quietly.

'It is my home, Inspector. My kids are all here. I can't move the whole family.'

'But he will come back later, and it isn't safe to be here.'

The woman shook her head, folded her arms, and bowed her chin.

'Does he have a firearm?' Inspector N asked.

'No, he does not.'

'Has he ever hit you, or threatened to hit you, with a heavy object?'

'No.'

'But he threw a chair at you this evening.'

She turned her face away, said nothing.

A teenaged girl appeared in the doorway. Inspector N looked up, watched the girl closely for a few moments, and then excused herself and went to speak to her. They disappeared together down a dark passageway and were gone for some time, perhaps ten minutes. In the silence that Inspector N left behind, her young colleague paced for a moment, and then sat down next to the woman. He could find nothing to say. They sat together at a corner of the big table, uncomfortably aware of one another's proximity.

When she returned, Inspector N walked up to the

woman and rested her arm on her shoulder. The girl stood watching them from the edge of the room.

'We have spoken to your Aunt Flora,' Inspector N said quietly. 'She is making space for all of you to sleep there tonight. And I phoned the station. They are bringing a vehicle big enough to transport you all to Aunt Flora.'

She paused, and the woman said nothing, just stared at her hands.

'I am sorry, sister,' Inspector N continued. 'I am only doing what you would be doing if you weren't in a state of shock.'

*

Some four hours later, at about two o'clock in the morning, Inspector N and her colleague responded to a complaint of disturbance of the peace at the other end of their jurisdiction. The woman in the big cold house was a mere memory now; we had responded to three calls since dropping her and her family at her Aunt Flora's home.

We came to a matchbox house that was silent and dark and appeared at first sight to be empty. Walking around the perimeter, Inspector N with her hands clasped behind her back, her partner beaming the light of his torch into the darkened windows, we stumbled across a woman sitting on a chair in the middle of the garden. She was very large, in her early fifties perhaps, and in the moonlight she appeared to be marooned there, as if she were alone and at sea. Inspector N walked up to her and greeted her. She grunted a vague response and attempted in vain to lift her hand. Next to her chair on the ground stood a half-empty bottle of brandy.

The young constable noticed the child first. She was watching us from the doorway to the backyard, leaning against the doorpost. She must have been about nine years old. The constable touched Inspector N's shoulder and pointed at the child. Inspector N raised an eyebrow, eyed the girl for a moment, then walked up to her and began to question her in a soft voice. The child said that her mom and dad sold liquor from this house, that they had left more than an hour ago, that the woman on the chair was a neighbour and a customer. She said, too, that her one-year-old brother was in the house sleeping.

'Show me,' Inspector N said urgently. 'Show me your brother.'

The two went into the house, and Inspector N came out alone a moment later. She moved briskly. The path she was hastily making pointed directly at the drunken woman, and from the anger in her footfalls it was clear that Inspector N planned to storm her. The constable and I both stood back and watched. Inspector N kicked the bottle of brandy hard, and it flew across the garden and smacked against the garden wall and shattered. The drunken woman flinched and rolled her eyes. Inspector N went down on her haunches into a crouch, grabbed a leg of the chair in each of her fists, and pulled the chair from beneath the drunken woman. She hit the ground hard, a terrible, dead thud, and then she was splayed and sprawling, her dress hitched above her knees. Inspector N hurled the chair against the garden wall, grabbed the drunken woman by the wrist, and began dragging her towards the van like a carcass.

'A baby!' Inspector N shouted. 'A one-year-old boy! Supervised by this old drunk!'

Her own children were no longer babies. One was 13, the other 16. She had told me some hours earlier that she resented Friday night shifts, that she did not like leaving her children alone at home throughout a weekend night. She always discovered some time during these shifts, she told me, an anxiety that had been throbbing beneath the surface of her consciousness all the while.

From her place on the ground, the drunken woman began to speak.

'Wo!' she said. 'Respect, sister! Respect!' Her voice was quite sober now, and surprisingly deep; in its indignation, an unmistakable authority.

Inspector N stopped in her tracks and let go of the woman's arm. She found herself staring down at a person perhaps ten years older than her, and a mother, no doubt, a person she had no right in any conceivable world to be dragging across the ground. She called to the young constable, in her voice a mixture of old anger and new shame, and together they helped the woman to her feet. I retrieved the chair. They took her to it, a shoulder under each of her armpits, and carefully settled her down.

I did not see Inspector N again, and never got to ask her to share her thoughts on what had happened. But I guess that her motherhood had leaked into her police work, and her police work into her motherhood, and while that was happening, she had done something that would later cause her to feel shame.

*

There is another way in which police officers get involved in the private lives of civilians.

In Reiger Park, on a Saturday night, the same night on which the second bleeding woman walks into the police station, Constables B and M are called to a complaint in a shack settlement at the edge of their jurisdiction. They are told over the radio that a father has demanded sex with his daughter.

They come to a home that is something between a formal house and a shack. The stand is large, it has a well-kept garden and a picket fence, but the house itself has clearly been made by its inhabitants; it is a mixture of wood, brick, concrete and iron.

The constables do not go inside. They hoot, and then watch while their engine idles.

A young woman appears from the house and comes up to the driver's seat window.

'Your dad's at it again?' Constable M asks.

She nods. 'He's gone out visiting,' she says matter-of-factly. 'When he comes back, he is going to want to fuck me.'

Constable M is drinking a Sprite. The young woman reaches into the car, grabs it from him, puts the straw carefully in her mouth, and sips.

'I'm jealous of your father,' Constable M says. 'Come to my place for the night. I'll protect you.'

She smiles to herself and hands Constable M his Sprite. His right hand is resting on his open window. With her index finger she traces a vein on the back of his hand.

And now her mother has come out of the house and is making her way to the car, and the girl takes her hand back and smiles. The cops are very respectful towards the mother. They write their cellphone number on a card and give it to her, and tell her that when her husband comes home she must phone them immediately, that they will drop whatever they are doing and come.

I discover during the course of the night that Constable B has a girlfriend in his jurisdiction, and that Constable M is looking for one. The following week, I join Constable T in his police station on the East Rand, the young man who once dressed as a chicken and has now bought a house in the suburbs. I find him on a quiet day. After lunch, we drive into the suburbs, pick his daughter up from school and take her home. Driving back into the township, Constable B looks at his watch, and thinks aloud that there is still an hour left before he must pick up his wife from work. He drives us back into his jurisdiction. We stop outside a house, and he hoots.

'My girlfriend's place,' he explains.

'You all seem to have girlfriends where you work,' I say. 'Is it a professional perk?'

'No,' he replies. 'It is giving back.'

'To whom?'

'To the community we police. These girls are very expensive. You have a girlfriend in your jurisdiction, it means you must leave at least a quarter of your salary here every month. It is redistribution, our gift to the township.'

Sibanda of the suburbs

A notable chunk of urban South Africa has been missing from this book: the suburbs, and, in particular, those that remain predominantly white. The story of the suburbs closes a circle, one that isn't always visible; it lies buried under the piles of debris and dross; one needs to rake and clean and scrub before one can make out its circumference. It is a circle that is finally closing here and now after three-and-a-half centuries, one that joins black and white in a common quest for security.

Until the late 1980s, or the early 1990s, perhaps, black and white South Africans lived in utterly different security universes. For generations, the South African state was able to underwrite the personal security of white people as well as any of the advanced industrial states in the northern hemisphere protected their own citizens. Rates of murder, armed robbery and grievous assault were as low among South African whites as they were among the middle classes of the gentlest and most inhabitable cities in the developed world.

As for the personal security of black people, the white state did little to underwrite that at all. It is quite extraordinary, for instance, that during the 1970s and 1980s, crimes like murder and armed robbery were doubling every five or six years in Soweto and Alexandra, and yet remained constant among whites. If you had told a Johannesburg suburbanite back in 1985 that her city's murder rate had quadrupled in the last 25 years, she would have stared at you blankly.

And so blacks and whites lived in parallel worlds. White people assumed that providing security was the role of the state. Black people knew that if they wanted security, they would have to acquire it themselves, whether in exchange for money, or neighbourliness, or ethnic or political solidarity.

That is the circle that is now closing. The transition to democracy has spread the condition of insecurity from black people to white. The South African state is no longer able to keep whites as safe as their suburban counterparts in Surrey and Amsterdam. White South Africans are now as heavily victimised by crimes like armed robbery as the middle classes of notoriously violent cities like Caracas, Bogotá and São Paulo.

And so, for the first time in the history of South African security, whites are starting to behave like blacks. Abandoning the state as a failed protector, they are beginning to organise personal protection on open markets, out of ethnic solidarity, out of neighbourliness. It is taking whites a long time to learn to behave this way; they have made some spectacularly false starts. Seeking one's own security does not come naturally if it isn't already in the blood. But coming it slowly is. And it is changing the face of the suburbs.

*

On an afternoon in late August 2007, I meet one of these new, white seekers of security. He is a cofounder of an initiative called Community Active Protection. The venue he has suggested is a coffee shop in the north-eastern Johannesburg suburb of Glenhazel, one of the city's few neighbourhoods that is almost entirely Jewish. That we are meeting here is integral to the essence of the story he is to tell me.

The early Jewish communities of turn-of-the-century Johannesburg clustered largely in neighbourhoods directly east of central Johannesburg: Doornfontein, Bertrams, Troyeville. As Johannesburg grew and Jews prospered, the community slowly spread northward across a long, narrow swathe of northeastern Johannesburg: Sydenham, Highlands North, Savoy, Norwood, Glenhazel, later, Linksfield.

Within this cluster of predominantly Jewish neighbourhoods, Glenhazel stands out. It is far wealthier than most of the other Jewish suburbs around it; among its residents are highly successful entrepreneurs, corporate executives, and flourishing professionals. And so it combines ethnic solidarity with large reserves of money. As a result, it is home to a dense concentration of Jewish social and cultural institutions. The Yeshiva College is here, as is the Selwyn Segal, a home for the handicapped. There are several facilities for the aged, as well as a rehab centre for substance abusers. In Sandringham, just outside the border of Glenhazel, there is a residential home for the mentally handicapped. And the South African chapter of the Chevrah Kadisha, the famous Jewish burial society and provider for the Jewish poor, is also here.

And so if there is to be a creative initiative to protect Jews in post-apartheid Johannesburg, this is the suburb from which it is bound to emerge. Just as in the context, say, of Alexandra, the best security initiative came from the ranks of old Alex families, their identities moulded by their role in the 1986 uprisings.

*

The man I meet at the Glenhazel coffee shop is about my age, I guess, late thirties. He wears a yarmulke on his head. He is not especially pleased to be talking to me. He exudes wariness. He is only here at all because I approached him via a trusted intermediary: South Africa's Chief Rabbi, Dr Warren Goldstein. Goldstein himself has been involved in this security initiative from the beginning.

And so my interlocutor says at the outset that he does not want me to use his name. I assent immediately and a little impatiently; I tell him that I am more interested in his story than in his identity. I shall call him B.

He gets immediately down to business; the story he tells of the security service he helped start is precise and clipped; it exudes at each of its junctures his lean, methodical intelligence.

'Some time in 2005,' B tells me, 'a couple of us started an initiative around crime in this neighbourhood. We knew that there was terrible crime in Glenhazel: home invasions, people held up at gunpoint going to *shul*, carjackings. But it was all anecdotal. We needed to collect information. So we opened up our networks. Eighty-five per cent of Glenhazel is Jewish. You can open a grapevine very quickly.

'We believe that our information gathering was successful; we believe that between 70 and 80 per cent of all contact crimes against Jewish people in this suburb were reported to us. So we came to a good understanding of what was happening.

'What we discovered was this. About 15 contact crimes were committed in Glenhazel every month. They were divided into three categories: carjackings, armed robberies in driveways, and home invasions.

'As far as robbers' methods of targeting victims are concerned,' B continued, 'we also picked up three patterns. The first was opportunistic: they'd drive by, or walk by, and come across a chance to rob. The second was reconnaissance crimes. The victim's asset was targeted, the crime planned. The third pattern was following: the victims were followed home in their cars, whether they were coming from the shopping centre, from the bank, from the airport.

'These three crimes – carjackings, armed robberies in driveways, and home invasions – were more or less equally dispersed: about a third each. So were the three methods of victim selection: opportunistic, reconnaissance, and following the car.

'The next thing we did, we gathered information on the anatomy of the actual crime itself: where it is initiated, where it moves to, how long it takes, how it ends. Carjackings and driveway robberies are two-minute crimes. It is very unusual for the whole crime to take any longer than that. Home invasions: it takes 15 seconds to enter the home, and then the rest of the crime takes place out of sight, behind those big walls everyone is building. Common to all three of these crimes is that they are initiated in public space, not private space. And public space was ownerless. Nobody was watching it. Nobody was protecting people in it. That was the problem.'

B is warming to his story now, for he is getting to the crux of the matter; while his careful, methodical manner remains unchanged, the rate of his speech quickens.

'So you have this situation where everyone is retreating into private space because of crime. They are building high walls around their homes. They are getting armed response. The walls are useless because the crimes are initiated outside them. And as for armed response, crime changed, but armed response didn't. Crime found the gaps between all the hard targets, and armed response kept hardening the targets. Armed response doesn't help. It just makes residents emotionally secure.

'We realised that if our initiative was going to be successful, we needed to forget the walls and concentrate on the point at which crime is initiated, and that is public space. We needed to disrupt crime in the public spaces where it starts.'

*

I put down my pen and offer B my private, silent congratulations. In the few minutes we have been together he has very quickly and deftly lifted the lid on the most spectacular hoax in the recent history of suburban security in South Africa.

Until the late apartheid years, Mtutuzeli Matshoba's blackjacks, together with an assortment of other security functionaries, kept the suburbs safe. While the police were still effective at controlling the movement of black people, no township gangster was free to plunder the white suburbs. The Hazels with whom Mtutu grew up in Mzimhlophe would mug black commuters on Soweto's transport routes, and hold up the drivers of township delivery vans in broad daylight. But the prospect of sauntering into the suburbs and robbing a white housewife of her car was utter fantasy.

In the mid-1980s, apartheid's spatial controls began to crumble. By the closing years of the decade, black people were pretty much free to move through South Africa's cities. For the township underworld, whose long pedigree had been invisible to white South Africa, the opening up of urban space heralded something of a revolution. The vastness of the new market promised by white neighbourhoods was almost too spectacular to imagine. A criminal culture whose appetite for commodities and for violence was legendary in the townships, arrived in the suburbs.

The residents of suburban South Africa have responded to this onslaught by recoiling. In a thousand shapes and forms, they have built castles and fortresses and buttresses, and ducked behind them. During the last decade and a half, South African suburbanites have probably done more than anyone on earth to reshape their cities into postures of self-defence. Where there was once open grassland around the periphery of Johannesburg, there is now one gated community after the next, standing there in tightly packed rows against the backdrop of the blond veld, most in the gaudy style of ersatz-Italian villas. The older middle-class suburbs closer to the centre of town are sealed off with palisade fences and road booms, and patrolled around the clock by armed security guards. Shopping malls have multiplied, and the old ones built in the 1970s and 1980s have doubled and trebled in size. Out in the suburbs, one sees office park after office park, each heavily access-controlled and ringed with electrically charged fences. I doubt whether there is another peacetime city on the face the planet where fear has been as powerful a factor in reshaping the built environment.

And yet if one looks at what has happened to our crime statistics over the last decade, this massive investment in safety has been a ruse. The crimes that the new barricades are meant to ward off have escalated. In the 1995/6 financial year, the police recorded just over 77 000 armed robberies per annum. In 2006/7, the figure stands at more than 126 000, an increase of almost two-thirds. In contrast, residential and business burglaries – in which empty homes and businesses are broken into and thus nobody is held up at gunpoint – have declined a little: 319 000 in 1995/6 down to 308 000 in 2006/07. The contrast is ominous. Eleven years ago, for every predator who held someone up at gunpoint, 4.1 empty homes or businesses were burgled. Today, the ratio has moved to 1:2.4. It seems that a decade of target hardening has simply spawned a generation of criminals prepared to use more violence.

Why did suburbanites throw so much money at so useless a remedy? Essentially because they acted as individual householders and not as communities, and because they live in a market economy. If your neighbours build higher walls, you must do so too. If your house is the only one around without a big wall, a predator who does reconnaissance in your street will target you. Next, your neighbour gets an electric fence, and you must follow. Then armed response. The security market thrives on the blind necessity of these cascading defences. It offers more and more, knowing that its clients must take and take. And so everyone keeps up with his neighbour, and, collectively, everyone is more exposed to violent crime than he was before the first walls came up.

*

B stands outside of this spiral and watches it. He can see what he does because his task is neither to fortify his house nor to make money: it is to protect a neighbourhood. He is a community leader; he lives in an ethnically homogenous suburb where ties are thick and bonds are strong. He is positioned to observe the whole.

'The task at hand,' B continues, 'is to disrupt criminal activity before it is initiated. How do you do that? The first thing we did is we began to mobilise the community. Everybody needs to be alert to what is unfamiliar. Everyone in the environment must immediately recognise anything that doesn't belong there.

'We built up this community alertness component over 18 months. It is a multi-tiered system of surveillance. There are block watches, neighbourhood watches. You see something, you report it. About 500 domestic workers have been on a training programme; 800 residents went through defensive training. Everyone here has been trained on how to spot what is out of place, and knows to report it.

'The next question is: where do you feed all this

surveillance? Initially, we had no private resources. We leveraged the South African Police Service and all private security firms with clients in the area. But we soon discovered that neither is designed to respond to civilian surveillance. Neither is designed to disrupt the preparation of criminal activity. Residents called the cops. The cops didn't respond. Residents got despondent.

'As for private security, they were all glad to assist. Armed response started responding to the new patterns of calls that were coming in. Their operational deployment started changing. They'd sit in the middle of the zone to decrease response time. But ultimately, they couldn't cater to what was needed because they are profit-driven. For instance, to save costs, they have one man in a vehicle. But one man is easily intimidated. He backs off when he is scared. He's not effective at disrupting criminal activity. Also, their mileage was restricted to save on petrol; there were things they weren't responding to.

'So we realised that community mobilisation wasn't enough. We built our own command and control centre, and we drew up specs for hiring our own security resource. The command and control centre is run by a non-profit company. Its board is filled with distinguished community members, people who are widely known and respected.'

As for the 'security resource', one sees it for oneself when driving through Glenhazel; clearly marked four-by-fours, two armed, black-uniformed men in the cab; a canvas canopy on the side of the road opposite the Yeshiva College, under the shade of which stand three men, each armed with a shotgun. The 'security resource' is staffed primarily by ex-military men, veterans of the Angolan war well represented among them. 'We were advised to recruit people who had seen combat,' B explained, 'people who would not stand down.'

Collectively, the residents of Glenhazel phone the command and control centre about 200 times a week to report suspicious activity. It is true that more than a thousand residents and domestic workers have now been trained to detect what is suspicious, but the criteria are in fact as plain as can be: two men or more, moving by car or on foot. For 'two men' one must of course read 'two black men'.

So when a resident sees two or more black men whom she does not know, she calls the command and control centre. A car is dispatched with its armed, ex-military men in the cab. What do they do when they arrive at the site of the suspicious black men? Here is where the initiative reaches the borders of legality and must take care. Stopping and searching is illegal; only the police can do that, and even then, only if they have reasonable grounds. The initiative has asked the local police station commander to spare personnel so that stopping-and-searching can be done by a police officer. He has said no. Any form of harassment is also illegal, of course. I have been informed that several people have come to the local police station complaining that they have been harassed by these patrols. Thus far, nobody has laid a criminal charge.

I put all of this to B.

'Our task is to disrupt,' he replies. 'To disrupt, you do not need to stop and search. You do not need to harass.

All you need to do is make yourself visible until they grow uncomfortable and leave. If two guys come into Glenhazel to commit a robbery, and armed men start following them around within minutes of their entering the suburb, they're not going to commit their crime. They're going to leave.'

I ask B whether I can accompany one the patrols for a day or two. He hesitates and thinks, then says he will get back to me. He never does. Clearly, he is more comfortable describing to me what the patrols do and don't do than having me see for myself.

Does it work? After a long period of denial, the local station commander finally admitted in September 2007 that since the launch of the initiative, armed robberies in Glenhazel had dropped by 66 per cent. According to B, who goes by what the Glenhazel community reports to him, rather than to the police, armed robbery has declined by 77 per cent.

*

When B and I sit down to meet, I refer to the initiative he represents as GAP – Glenhazel Active Protection.

'That's not right,' he says immediately.

'But that's what it says on the sides of your patrol cars,' I reply.

'Still, I'd rather you referred to it as CAP – Community Active Protection. We might have started in Glenhazel, but we have gone well beyond that now. We're in Sydenham, in Savoy and Waverley. And that's just the start.'

There is in fact a very similar initiative in the upmarket suburb in Sandhurst. But a rift has grown between them and B's people. B chooses not to mention it.

'We plan to start on the other side of the highway, too,' he continues. 'In Victory Park; around there. Where do you live?'

'In Killarney.'

'Well, it's coming to you. It's on the way. There's an initiative in Houghton, just around the corner from you. This is going to be citywide. This can work everywhere.'

The question of the initiative's boundaries is clearly a sensitive subject. B is at pains to declare that he is a patriotic South African. He tells me over and again that Community Active Protection has not established itself in competition with the police. It is a support, an auxiliary: it must work in partnership with the police. And it is not, he tells me several times, an insular, tribal initiative. Yes, it was started in, and by members of, the Jewish community. But it is not exclusively Jewish property. It is a resource that can be replicated across the suburbs, and perhaps in townships too.

B may well believe that, but I don't, and I tell him so. In whichever direction you care to move, the further from Glenhazel the initiative spreads, the scarcer one or the other of its two crucial ingredients become: money and ethnic solidarity.

The initiative does not come cheap: the spanking new four-by-fours, the premium salaries of the soldiers, the maintenance of a continual capacity to respond to all 200 calls a week. Glenhazel residents pay R500 a month for this service. That is not a big sacrifice for an upper-middle-class household; it is for many of the households due west of Glenhazel. B tells me that Glenhazel residents' fees are already subsidising one of the areas that has recently joined the initiative. For how long this redistribution will be tolerated one does not know.

To be fair, though, money ought not to be decisive. Chasing black men out of neighbourhoods need not be an expensive business. The residents of Glenhazel are buying the very best, but there are many cheaper, rougher ways, ways that substitute residents' time for their money.

Ethnic solidarity is probably the more important of the two ingredients. Tightly knit neighbourhoods like Glenhazel trust the people delegated to manage the initiative, especially when they are carefully chosen for their goodwill and their prestige. As for resident participation, free-riding is shameful when bonds are strong. One pays a heavy price for refusing to put in one's due. And when neighbours are not strangers, common security becomes a standard fare of table talk, and active participation in the initiative regenerates itself.

A friend of mine is a photographer. In early 2007 he was granted permission to take pictures of the black-clad men who stand with shotguns opposite the Yeshiva. He went early on a Friday evening, when residents were walking to *shul*. Within minutes of his arrival, he found himself surrounded by four men, all young, all Jewish residents, all in *shul*-going suits. He explained that he had permission, that he was no stranger, that he had an uncle and cousins in the neighbourhood, that he meant no harm.

'We don't care who you are,' one of them replied. 'No pictures.'

'I'm doing no harm,' my friend repeated.

'You are. Pictures kill. Pictures started the intifada. Pictures inspired suicide bombers to kill Jews. Now get out of here.'

It is not everywhere in Johannesburg that one finds such ample reserves of community; this is a village that has hunkered down. Across town, there are suburbs like Parkview, Parktown North, Greenside. They are easily as wealthy as Glenhazel, but they are not communities, and they never will be. They will never produce a protection service anything like this one.

*

Ultimately, security initiatives like B's start creating the sort of copycat pressures that the walls and the electric fences did during the 1990s. Robbers scared off from well-guarded places like Glenhazel will rob elsewhere. Suburbanites who do not want to attract these displaced crimes will try as well as they can to do as Glenhazel does. Some will succeed; their initiatives will not look quite like Glenhazel's; each neighbourhood builds something after its own fashion, something that reflects what it is. Neighbourhoods that have hosted the same families across the generations will do well, as will ethnic clusters, like the pockets of Portuguese-South African settlements in the east and the south, and like the middle-class Muslim families currently moving back into inner-city Fordsburg in large numbers.

Those not blessed with this asset of neighbourly solidarity will suffer. To live in a suburb where the resources of self-organisation are poor is to be less secure.

In this regard, white urban South Africa is truly beginning to resemble black urban South Africa. Glenhazel, I guess, approximates Newclare of the early 1950s: predominantly Basotho, village-like in its solidarity and its wariness of other Johannesburg residents, admirably proficient in organising self-protection. Ephraim, the Alexandra murder victim, and his neighbours also have their equivalents in middle-class Johannesburg; suburbs in which few know their neighbours, let alone trust them, and the best form of self-protection is to recoil from the city and to keep one's head down.

*

I phone Rabbi Goldstein to thank him for facilitating my meeting with B. He tells me it is of the utmost importance that the initiative moves beyond the suburbs and finds a place in Alexandra. So many of the people who come to the northern suburbs to rob are from Alex. It is from there that the guns come, and it is to there that the hijacked cars and stolen goods are spirited. Surely, he thinks aloud, there is a wide space of mutual interest between Alex residents and suburbanites; these young men with guns – surely the fear of them is something that is mutually shared. He says that he has begun to send out feelers, and is fast getting the impression that the politics of civic life in Alex is monumentally complicated. But he is keeping his head down; he is determined to succeed.

The rabbi's instincts are surely right. It is a grim business watching suburbs build moats and drawbridges around themselves, bouncing the problem back into the rest of the city. It is better to climb to a vantage point from where one can see the whole, and act upon problems that afflict the whole.

I suspect, though, that the rabbi's goals for Alexandra will prove elusive until policing starts working far better. There is no substitute for a state agency charged with policing people and investigating crimes. When it is not working well enough, people will inevitably build their moats and drawbridges. Dan Sibanda will protect himself and his kind after his own fashion, B after his, Ephraim's neighbours after theirs. The city is a motley patchwork of self-defence.

*

This book began with an account of two constables on patrol, and so it shall end.

Glenhazel falls into the Sandringham police station precinct. It consists of northeastern Johannesburg's sizeable rump, beginning with Sandringham and Glenhazel at its southwestern extremity, extending through the suburbs of Lombardy East and West, Bramley Gardens and Kew, skirting around Alexandra, and incorporating an area of peri-urban smallholdings east of the N3 motorway. On an ordinary day, the task of policing this massive jurisdiction falls upon two patrol cars.

The constables I accompany on a weekday patrol are not especially pleased to have me with them. They quickly ascertain that I understand little Tswana, and begin speaking it exclusively. They answer my questions curtly and continue talking to one another.

We pull up at a traffic light, and an old man approaches the window. He is holding a homemade placard. 'In the name of God,' it reads, 'take pity on me.'

One of the constables rolls down his window.

'What do you want?' he asks the old man in Tswana.

'*Ngilambile.*' 'I'm hungry,' the beggar replies in Zulu.

'Have you ever prayed?' the constable asks earnestly.

'Every day.'

'Well, then, pray now. Let's see how you pray.'

The old man closes his eyes and begins to speak: 'Our father in heaven …'

Both constables roar with laugher. One slaps his thigh, the other the steering wheel. We drive off, the tyres squealing with cheap pleasure, the old man still at prayer in the rear-view mirror.

*

Shortly after five o'clock in the afternoon, the noise and fumes of rush hour swirling about us, the two constables set up a vehicle checkpoint on their jurisdiction's busiest road. One of them inspects my bulletproof vest carefully, shakes his head and adjusts it, and tells me to hang back from the roadside and remain in the vicinity of the van.

'If we flag down a car carrying an illegal weapon,' he says, 'they will either speed away or shoot. It is better if you are not standing close to us.'

The first vehicle to stop at our checkpoint is a police patrol van marked Sandringham. It parks in the tramlines, and both of its occupants get out and come to us.

'What are you doing?' one of them asks in Zulu.

'A vehicle check point.'

'Why?'

'Because we have a researcher with us. He wants to see police work.'

'You've chosen the main road into Alex.'

'So?'

'You want to get shot for a researcher?'

The two cops from the other van laugh ungenerously, stroll back to their vehicle, and drive off.

Once they have gone, the two constables break into heated discussion. Again, I do not understand their Tswana, but they are clearly angry with one another. Whatever their dispute, it is quickly resolved. We get back into the car, turn into a narrow side street, mount the pavement, and reconvene at the vehicle checkpoint.

It does not take long to see that the cops have chosen their position with care. We are about 50 metres beyond a stop sign, from which we are clearly visible. If the driver of a car does not wish to encounter the two constables, he need simply turn left at the stop sign down a quiet suburban street, loop around us, and continue with his journey as planned.

And that is precisely what four out of five drivers approaching the stop sign do. The fifth is the unlucky one who was distracted; she was talking on her cellphone, or daydreaming, or lost in anxiety. She is certainly not armed.

The constables are not bothered now about where I stand, for there is no danger to any of us. I join them up front, at the road's edge.

'Tell me about your experiences of GAP,' I ask one of the constables.

'What's that?' he replies.

'Glenhazel Active Protection.'

'Oh, the private security. They all do their bit.'

'But each is a little different from the next,' I say. 'GAP, for instance, is not quite the same as Chubb.'

'No,' he replies. 'They are the same. Whether in Glenhazel, or Lombardy East, or wherever, private security is private security.'

'How long have you been at Sandringham?' I ask.

'This is my fifth year.'

Another dozy driver has strayed towards our checkpoint, and the constable steps into the street and flags her down. I am left with my thoughts. About a year ago, an entire suburb in this man's jurisdiction turned from the police and erected a substitute agency. To all intents and purposes, several thousand people here have severed their relationship with the South African Police Service. He appears not to have noticed.

Notes

20. *'Citizens must be submitted to authority …'* Cited in Jean-Paul Brodeur, *A Treatise on Policing* (Toronto, forthcoming).

65. *'patrol the streets, to disarm people …'* Gary Kynoch, *We are Fighting the World: A History of the Marashea Gangs in South Africa, 1947–1999* (Pietermaritzburg and Athens, Ohio, 2005) p 95.

66. *'these migrants were able to engage …'* Kynoch, *We are Fighting the World*, p 110.

66. *'The Russians retaliated …'* Kynoch, *We are Fighting the World*, p 95.

90. *'recognise the importance of crime as a popular grievance …'* Clive Glaser, *Bo-Tsotsi: The Youth Gangs of Soweto, 1935–1976* (Johannesburg and Oxford, 2000), p 177.

105. *'None of their friends and family had a better house …'* Adam Ashforth, *Witchcraft, Violence, and Democracy in South Africa* (Chicago, 2005), p 29.

Further reading

The thesis that the consent of citizens to be policed is a precondition of policing, so central to the main argument of this book, was developed with admirable lucidity by Egon Bittner in the 1960s and 1970s. See, in particular, Egon Bittner, *Aspects of Police Work* (Boston, 1990). I also profited a great deal by reading chapters from Jean-Paul Brodeur's as yet unpublished new work on policing; both his exegesis and critique of Bittner's work were extremely valuable to me. See Jean-Paul Brodeur, *A Treatise on Policing* (Toronto, forthcoming). See also Jean-Paul Brodeur, 'An Encounter with Egon Bittner', in *Crime, Law and Social Change* (Volume 48, Numbers 3–5, December 2007).

South African historians have in general written far too sparingly on the history of security in black urban life, but there are a few gems which helped me enormously with the historical aspects of this book, especially, Clive Glaser, *Bo-Tsotsi: The Youth Gangs of Soweto, 1935–1976* (Johannesburg and Oxford, 2000); and Gary Kynoch, *We Are Fighting the World: A History of the Marashea Gangs in South Africa, 1947–1999* (Pietermaritzburg and Athens, Ohio, 2005). See, also, Adam Ashforth: *Witchcraft, Violence, and Democracy in South Africa* (Chicago, 2005); Belinda Bozzoli, *Theatres of Struggle and the End of Apartheid* (Ohio and Johannesburg, 2004);

Philip Bonner and Lauren Segal, *Soweto: A History* (Johannesburg, 1998); Ellen Hellman, *Rooiyard: A Sociological Survey of an Urban Native Slum Yard* (Manchester, 1969); Charles van Onselen, *New Babylon, New Nineveh* (Johannesburg, 2002); Charles van Onselen, *The Small Matter of a Horse: The Life of Nongoloza Mathebula, 1869–1948* (Pretoria, 1984).

The following memoirs and autobiographical fictions also helped shape my thoughts about urban security under apartheid: Hugh Masekela and D Michael Cheers, *Still Grazing: The Musical Journey of Hugh Masekela* (New York, 2004); Godfrey Moloi, *My Life: The Godfather of Soweto* (Johannesburg, 1989); Mtutuzeli Matshoba, *Call Me Not a Man* (Johannesburg, 1979); Can Themba, *Will to Die* (London, 1972); Sipho Sepamla, *A Ride on the Whirlwind* (Johannesburg, 1981); Ezekiel Mphahlele, *Down Second Avenue* (London, 1959).

On policing and the transition to democracy in South Africa I was helped a great deal by Mark Shaw's indispensable book, *Crime and Policing in Post-Apartheid South Africa: Transforming Under Fire* (Bloomington and Johannesburg, 2002).

While writing this book I had constantly in my mind several other books on policing and justice in contemporary South Africa; each played, in turn, the functions of sparring partner, reference, and thought provoker. See Antony Altbeker, *The Dirty Work of Democracy: A Year on the Streets with the SAPS* (Johannesburg, 2005); Antony Altbeker, *A Country at War with Itself* (Johannesburg, 2007); Diana Gordon, *Transformation and Trouble: Crime, Justice, and Participation in Democratic South Africa* (Ann Arbor, 2006); Jennifer Wood and

Clifford Shearing, *Imagining Security* (London, 2006); Les Johnston and Clifford Shearing, *Governing Security: Explorations in Policing and Justice* (London, 2002).